T0141079

Deep Economy

Caring for Ecology, Humanity and Religion

Deep Economy

Caring for Ecology,
Humanity and Religion

Hans Dirk van Hoogstraten

James Clarke & Co
Cambridge

First Published in 2001 by:
James Clarke & Co
P.O. Box 60
Cambridge
CB1 2NT
England

e-mail: **publishing@jamesclarke.co.uk**
website: **http://www.jamesclarke.co.uk**

0 227 67966 0 hardback
0 227 67965 2 paperback

British Library Cataloguing in Publication Data:
A catalogue record is available from the British Library.

© Hans D. van Hoogstraten, 2001

All rights reserved. No part of this publication may be reproduced,
stored in a retrieval system or transmitted in any form or by any means,
electronic, mechanical, photocopying, recording or otherwise,
without the prior permission in writing of the Publisher.

Printed in the United Kingdom by
IBT Global

CONTENTS

To Dorine,
expert in architecture
and beyond. . . .

Acknowledgments

TO WRITE A BOOK, one needs, among many other things, concentration and inspiration. Several people and institutions have helped me concentrate by taking burdens off my shoulders, and have inspired me by providing new burdens. Nijmegen University stimulated my research in the fields of philosophy, religion, and economy. During the 1990s, I was granted several periods of study and teaching abroad, such as my three-month stay at the Divinity School of The University of Chicago, and a five-month teaching exchange with Professor Louke van Wensveen at Los Angeles Loyola Marymount University. I want to express my gratitude to Loyola Marymount University, whose generosity enabled me to work on the book's text with an excellent editor: Lauri Radin, M.A. I also enjoyed attending several American Academy of Religion Annual Meetings, and numerous Conferences of the International Bonhoeffer Society. All these activities, including countless opportunities to lecture on and discuss relevant themes concerning (social) ethics, contributed to the development of my thought, presented in this book.

Certain ideas in this book were published earlier in my Dutch book *Geld en geest: over milieu-ethiek* (Baarn: Uitgeverij Ten Have, 1993). I want to express my appreciation for the publisher's consent to include these here.

Among my friends and colleagues who read the manuscript, commented on it, and wrote endorsements, I wish to mention: Ulrich Duchrow (Heidelberg, Germany), Larry Rasmussen (New York), John B. Cobb, Jr. (Claremont), and Gerard Heather (San Francisco). Their fine and precise work is much appreciated.

My son Diederik, graduate of Columbia University's School of Journalism (New York) translated and edited parts of the text and was a faithful reader and discussant. Several others are part of the process of developing thoughts, accents, and connections. As a representative of my friends and partners in discussion, I want to mention Pieter Arends, psychiatrist and philosopher. My students are represented honorably by Eward Postma, my outstanding student-assistant.

The most important person, however, is my partner for life, who happens to be a discussant for life as well: Mies van Hoogstraten-Dorsman.

Nijmegen
The Netherlands
summer 2001

This book's core issues and related subjects are discussed in addresses and seminars of the Amsterdam Institute of Deep Economy and Ethics (IDEE). Discussions on line are also possible – see www.deepeconomy.net.

INTRODUCTION

HISTORICALLY, SOME VERSION OF the Word of God has been the basis for ethical reflection and moral action. God's Word has always been mediated by an official person or institution. In recent times, however, other voices have risen to the audible level. The deep ecology movement in the United States is a particularly strong new voice.[1] This group promotes the concept of reality as biospheric wholeness. As one of numerous species, humans have a responsibility to behave in a manner that does not disturb food chains and the ecological system at large. From the immediate experience of the wilderness, deep ecology draws fundamental conclusions about human thought and behavior. Deep ecology's ethical imperative calls for the adaptation of humankind's personal, political, and economic actions to the reality of nature. Thus, nature is given the status of a moral standard, which could be called divine.

Deep economy is not just a play on words—it is the area that we must address to find the root causes and final solutions to our present ecological crisis. We must examine human motives, influences, and manipulation both on a surface level and on a deep level. Economy is inextricably linked to modern environmental problems. Many environmental ethicists believe that economic actions are the fundamental agent in environmental problems and warn that profound changes in economic thinking and acting must be made.[2] In spite of their diagnosis, however, nothing changes. The reason for this perpetual inaction is the habitual denial and neglect of the economy's 'deep' character. Is our inability to change the economic system really a deep-rooted unwillingness to tamper with the system?

This book outlines the development of modern Western commercial society during the last few centuries. It shows that numerous elements from the period known as the Enlightenment were incorporated to establish our modern 'free market.' Economy's deep character is the result of the working combination of the dominant worldview, psychology, technology, science, mathematics, and sociopolitical reorganization within a process of historical development. This entire process is dominated by the idea of progression and its influence on human thought and by the idea that human beings' position in nature should not be underestimated. In fact, the character and position of modern society's economy can be identified as a worldview and a religion. Thus, the deep character of economy must be acknowledged and examined.

Economy is an important aspect of Western enlightenment, which was a lengthy historical process that can be roughly localized in two prominent 'waves.' The first wave of enlightenment occurred in Greece during the fourth century B.C. The second wave occurred in western Europe during the eighteenth century.[3] Both waves established new scientific and social paradigms by incorporating useful features of previous systems and discarding features that were not useful. The economic aspect of each wave of enlightenment was initiated in the name of righteousness and in the interest of human freedom and autonomy. The economy has historically provided the necessary means for the improvement of humanity's position on earth. Thus, the modern pursuit of an independent, autonomous, and prosperous position in society and in nature has economic roots. In practice, however, this pursuit means defending one's own and protecting oneself against real and imagined threats from both human and natural forces. This modern stance of protecting oneself from the *other* has led to deadly and costly consequences for humans and nature alike.

The gravity of the current world situation calls for a third wave of enlightenment. Why again *enlightenment*? Because this concept offers previously unknown possibilities of connecting economy, reason, and religion. A new kind of enlightenment could offer a new, unexpected sequence and relation of these aspects. This, however, cannot happen without reconsidering these concepts' self-evident antagonistic relation. In a long historical process, economy was emancipated from religion, and reason tried to overcome religion. But as Clifford Geertz claimed, enlightenment's critical attitude toward religion works twofold. Religion is unmasked as pure fiction, and religion is also discovered as a real power, be it projection, imagination, or whatever.[4] The third wave of enlightenment that I am advocating will reveal economy's religious character. Liberation from this powerful ideology is definitely possible, just as the autonomous, enlightened person is entitled and able to free himself or herself from all sorts of projection. The difference between economy and projection as religious phenomenon is economy's power, rooted in a social, collective, mostly unconscious, belief.

What kind of therapy should be applied? The current trend of looking to premodern societal structures is a dead-end solution for modern society. Our efforts should be focused on developing a new paradigm, a paradigm that will bring about the third wave of enlightenment. In this book, I seek to define what is important and necessary for the development of a new paradigm. I begin by discussing the historical processes and achievements of the first two waves of enlightenment. We must understand the roots and the important historical shifts in order to understand our society today. Once these processes are understood, I identify our current system's strong and weak points, as well as its promising and destructive aspects. At this informed point, we will be in a position to make responsible decisions about the type of new paradigm that must be developed in order for humanity and nature to have a future.

Ian Barbour concentrates on this phenomenon in his two-volume *Gifford Lectures*.[5] Barbour states that the construction of a new paradigm must provide a "frame of reference for members in society [that respects] the collection of norms, beliefs, values, habits and survival rules." This provision is requisite, because there is "a mental image of social reality that guides behavior and expectations." Thus, our task is to establish a paradigm shift that features a "radical transformation of the scientific imagination," and at the same time, we must reinterpret existing cultural data in new ways.[6]

A new paradigm presupposes that there is an already functioning paradigm that is in serious need of improvement or radical transformation. Agreement among affected community members is a condition for the establishment of a new paradigm. Certainly, evidence and emphasis in the spheres of science, economy, politics, and morals are powerful and important enough to bring about a new paradigm. Yet in practice, it is difficult to get everyone involved, and once everyone is involved, it is difficult to get everyone to agree at the same time.

The need for a new paradigm that will change political, economic, and personal behavior and practices is well known. A virtual sea of literature exists on environmental problems, overpopulation, and the disparity of wealth in the world, and these are just a few of today's major global problems.[7] This book does not aim to add more data to the already existing profusion. Instead, this book explores the character of modern economy, including economy's ideological and scientific roots and its evolution in the process of Western enlightenment's development and shifts. This book seeks to identify the fixed rules and practices that threaten life today so that they can be dismantled, and it seeks to define the conditions that must be met for the establishment of a new paradigm for the third wave of enlightenment.

CONTENT OVERVIEW

The book begins by using a historical method to analyze the ethical and economic thought of Aristotle and Adam Smith. These two thinkers are presented as the main representatives of the first and second waves of enlightenment, respectively. This discussion clarifies why it is necessary to speak of *deep economy*. The ideas of the late Dutch scholar Arend van Leeuwen, with whom I cooperated on an economic theology project, are also presented in chapter 1.[8]

Chapter 2 focuses on the work of Herman Daly and John Cobb. Daly and Cobb discuss the fields of economy, theology, and philosophy, and they try to establish a critical connection with deep ecology.⁹ They have made valuable contributions to their field but have avoided deep economic discussion.

Chapter 3 identifies and discusses Western culture's central problem: dualism. The argument parallels deep ecology's claim of reality's connectedness. The difference here, however, is that I look toward human history rather than toward ontological cosmic perspectives. This chapter outlines the ancient Greek and Hebrew roots of Christian culture. It traces the dualism between humanity and nature, which is now commonplace in Christian culture, to the intermingling of these two ancient cultures. A hermeneutical method of literary interpretation is employed to reveal the difference between the original texts in their proper contexts and the Christian interpretation of the ancient texts.

In chapter 4, the overpopulation crisis is put where it belongs: in the broader discussion of the balance between life and death. This chapter draws on Latin American liberation theology, particularly the work of Costa Rican theologian and economist Franz Hinkelammert. Historical and modern practices of sacrificing life for success are discussed in depth. An effort is made to identify power's effectiveness in social elite's control of the hidden, mysterious origins of both new life and wealth. The chapter ends with a discussion on feminist perspectives and contributions, including those of Betsy Hartmann, to the population debate as a matter of women's reproductive rights.¹⁰

Chapter 5 applies the meaning and the role of money to history and nature and examines their mutual relation. Money as a deep economic factor is present throughout the book, but here, its function in the fields of politics, psychology, aesthetics, and ethics becomes crystal clear. I show how the age-old problems of debt and guilt have grown from personal issues to critical matters of global concern. The power relations between the First World and the Second and Third Worlds are explained in terms of credit and debt, life and death. Here, again, I discuss feminist perspectives, confronting ecofeminist approaches of reality with some of my inquiry's results.

The epilogue summarizes the book's different themes under the heading 'wrong connection.' Trinity's metaphorical power is discussed one more time, but this time it is strongly connected to Levinas's philosophy of the face of the other and Bonhoeffer's theology of being there for others. Both thinkers pretend that their accent is transcendent in character, which opens new, *post*-postmodern perspectives for society, community, and the individual. The final conclusion could be stated as "the belief's proof is in the action" (or should we say 'the absense of action'?).

The development of a new paradigm must take place in the First World, which has the power and the resources to make substantial widespread changes.

Poor countries and the growing numbers of refugees must awaken the consciousness of the First World. People who are aware of the tradition that they are in, as opposed to the awareness of one's so-called natural community, could join hands and create a true transforming power. Jewish thought and experience come to the foreground here, particularly in the work of French philosopher Emmanuel Levinas. His philosophy of the other applies to economic and ecological relations in the modern world. Levinas provides us with an ethical theory that we can use in facing today's problems. It answers the critical problems of exploitation and exclusion. Deep economic research lifts buried truths, such as the exploitative basis of modern economy's structure, and exposes them to the daylight of enlightenment's reasonable ethics.

It is my hope that the debate between deep ecology and deep economy will generate healthy new perspectives—the type that I can hardly even dream about today. However, everyone should be aware that major developments, such as new paradigms and waves of enlightenment, necessitate broad discussion and action in many different areas and on many different fields. It is the purpose of this book to contribute to these discussions and this development.

NOTES

1. See Bill Devall and George Sessions, *Deep Ecology: Living as if Nature Mattered* (Salt Lake City: Peregrine Smith Books, 1985).
2. See Herman E. Daly and John B. Cobb Jr., *For the Common Good: Redirecting the Economy Toward Community, the Environment, and a Sustainable Future* (1989; Boston: Beacon Press, 1994).
3. See Hans-George Gadamer, "Muthos and Wissenschaft" (Myth and Science), in *Christlicher Glaube in Moderner Gesellschaft* (Freiburg: Herder, 1982), pp. 6–42.
4. Clifford Geertz, *The Interpretation of Cultures* (New York: Basic Books, 1973), p. 93, speaks of religion as a model of and a model for human beings. See also Howard Eilberg-Schwartz, *God's Phallus* (Boston: Beacon Press, 1994), pp. 16f., who discusses the Freudian and the feminist approaches to religion.
5. Ian Barbour, *The Gifford Lectures*, vol. 1, *Religion in an Age of Science;* vol. 2, *Ethics in an Age of Technology* (San Francisco: Harper, 1990, 1993).
6. Barbour, *Gifford Lectures*, 2: 258. See also his description of a paradigm (1: 51): "Thomas Kuhn defined paradigms as 'standard examples of scientific work that embody a set of conceptual and methodological assumptions.' In the postscript of the second edition of his book he distinguished several features that he had previously treated together: a research tradition, the key historical examples through which the tradition is transmitted, and the metaphysical assumptions implicit in the fundamental concepts of the tradition. . . . A paradigm provides an ongoing research community with a framework for 'normal science.' Science education is an initiation into the habits of thought presented in standard texts and into the practices of established scientists."
7. See the indexes and endnotes in Barbour's volumes.

8. See Arend Th. van Leeuwen, *De nacht van het kapitaal: Door het oerwoud van de economie naar de bronnen van de burgerlijke religie* (Capital's Night: Throughout Economy's Jungle in Search of Civil Religion's Sources) (Nijmegen: SUN, 1985).
9. Daly and Cobb, *For the Common Good*, p. 9.
10. Betsy Hartmann, *Reproductive Rights and Wrongs: The Global Politics of Population Control and Contraceptive Choice* (New York: Harper and Row, 1987).

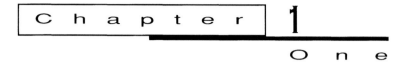

DEEP ECOLOGY
AND DEEP ECONOMY

WHY NOT COMPARE OUR secularized, technocratic, anthropocentric social system to a religion, particularly Christianity? The connection lies in
modern Western culture's identity as Christianity's heir. To a (post)modern
audience, it may seem strange that I explain and illustrate the domination of
the market-oriented economy by means of religious symbols and metaphors.
Let me explain my reasons for choosing what I believe to be a powerful
scheme of interpretation.

If we presuppose that people are always and everywhere looking for symbols, it makes sense to pay attention to historically dominant symbols with
multiapplicable effects. Trinity is one such symbol—maybe the most important one of Western culture—and it is impossible to neglect for people with
historical consciousness.[1]

Trinity presumes incarnation: God revealing himself in human flesh.[2] In
an era of mythical imagination, this revelation was of utmost importance:
humans are not totally submitted to a totalitarian deity and its oppressive
representatives. The incarnational structure is oriented to the biblical concept of the covenant and means liberation from arbitrary power relations—
at least potentially. Of course, the principle of incarnation can also be interpreted
the opposite way: people and institutions considering themselves the legiti-
mate successors of God the Son. This is what actually happened in church
and society and what takes special shape in today's market-oriented society
as I demonstrate later.

Before discussing human images of a divine being and their ideological
use, we should concentrate on the Trinity metaphor's content. In this metaphor, one of the most intimate human relations is projected onto God's being
and his activity toward humanity. The Father represents origins and tradition,
including traditional values and principles. God the Father shares his wisdom—
not keeping it for himself, hidden in heaven, but sharing it with humans,

1

revealed on earth. So fathership is lifted to a heavenly, transcendent level.[3]

The Christian father-son relation has its roots in the history of Israel. Having been called from slavery to freedom, Israel is declared the Son of God.[4] Thus, supporting liberating human acts, God is the symbol of human freedom. But the Father also provides the law as a condition to stay in a free and just society. The genuine Son of God is obedient to the law and to the commandments. The Father inspires the Son, sending him the Spirit (*Ruach*) as a creative, life-sustaining act.

The Israelites are expected to function as God's sons. Christianity personalized the Son in Jesus Christ, providing him with a divine as well as a human nature. This historically new scheme of divine-human relations bridges heaven and earth, making the Son into a heavenly king as well as an earthly gestalt: the Christian community is called the 'body of Christ,' in which doctrine and morality go hand in hand.[5]

During the first centuries of Christianity's religious-cultural development, the concepts of Father, Son, and Spirit are rather empty because of their transcendent character. This emptiness entitled people to consider themselves God's (the Father's) representatives (Son and Spirit), declaring official offices as powerful saving institutes.[6] For ages, it seemed hardly possible to separate church and state, altar and throne, pope and emperor ('political theology'). Only in modern times has God's position been questioned. People became aware that they didn't need the transcendent legitimation anymore.

My proposal in this book is to see what happens if we reapply the divine metaphors to our secularized social system, reclaiming the metaphor's analytical power. My presupposition is suspicion: I simply don't believe that an important, life-ruling religious concept such as Trinity, including its social and spiritual interpretations, suddenly vanishes when traditional religion decreases and the metaphor of the Trinity disappears from public discourse. In other words, there *is* a God who still rules the world—even if nobody calls him that.[7] God the Father as the creative, life-guarding, and life-providing power *is* working, be it under a false name. Applying old-time religious concepts to our technological, market-oriented society is the first step in our analysis. The second one has to do with this God's character, comparing the technological exchange society's 'divine face' with important characteristics of the biblical God, especially as interpreted by certain Jewish thinkers such as Emmanuel Levinas. Importantly, Christian thinkers such as German Dietrich Bonhoeffer, inspired by and trained in the Judaic religious tradition, analyzed God's character in similar ways.[8]

So post-Christian society bears a doctrinal burden. In terms of linguistics and imagination, doctrines and convictions have changed, but some of them still reign in a hidden way. Many treatises on environmental ethics and on deep ecology analyze these remnants. Yet these writings hardly ever offer an analysis of the deep religious character of economic theory and practice. A striking exception is Herman Daly and John Cobb Jr.'s *For the Common Good*.[9] They make a serious effort to reveal the free-market economy's ideology and practices.

Analogous to 'deep ecology,' I propose a new term: 'deep economy.' This characterization strikes the roots of the economy's current dominant position in society. Politics and ethics both depend on this structural framework. Political and moral theory and practical action have been developing in a condition of restrained autonomy. People believe that if they obey legal and moral laws, protection and prosperity belong to them. We must ask why they believe these political-economic promises.

We can formulate the answer in religious language. In terms of the Holy Trinity, we could say that the place of God the Father is taken by the economy, the place of the Son by politics, and the place of the Holy Spirit by ethics. The Son (politics, government) executes the will of the Father (economics, free market), and they both send the Spirit (ethics, ideology) to the people. The Spirit provides social directives and ethical rules.

As ironic and blasphemous as this may seem, the old Christian symbol of the Trinity aptly outlines the background of the (eventually) fatal thinking and acting of contemporary humanity. Morally, it seems that people can decide for themselves the proper way to live; within the boundaries of the norms and behavioral rules, there are plenty of free choices. Yet people cannot permit themselves to deny the will of the Father (respect for the rules and relations of the economic sphere). This producing and distributing, supplying and demanding Father is the liberating and prosperity-giving God. But he is also the avenging God who levies punishment when his laws are not respected. It concerns an ultimate gravity, a metaphor that signifies that matters have taken a grave turn.

The old metaphor of the divine Father is disputed on several sides. Feminist theologians, such as Sally McFague and Anne Primavesi, have applied convincing and constructive arguments concerning the negative impact of the father image, and they have proposed possibilities for new models.[10] My argument is different. In our secular 'economic religion,' we must begin by acknowledging and coping with the presence of a divine power. There is a striking resemblance between my analysis and the feminist critics' attack on patriarchal, hierarchical relations in the Christian culture. The religious symbol's role in the establishment and continuation of traditional power relations is particularly evident.

The hierarchy within the Trinitarian formula indicates the distribution and the exercise of power. The historical interpretation of the Trinity reveals how the metaphors are connected, how the Son and the Spirit realize the will of the Father. All of them are, theologically and economically, engaged in the 'economy of salvation.' Much research has yet to be done on the different meanings of this economy of salvation. My goal is to provide evidence for this unexplored field.

Moral philosophy's role in the establishment of modern free-market economy cannot be overestimated. A close selective reading of Adam Smith's theory makes this quite clear. Thus, it is appropriate to award ethics a divine character: the Spirit of Morality.

In Western thought, the emancipation of Spirit and morals from ecclesiastical power has been a long process, which started with Joachim of Fiore (1131–1202). He was the first one to use the Trinitarian symbol, expressing a heretical view of history. The era of the Spirit is connected to world history and is thus loosened from the church's claim of being the Spirit's only true representative and dwelling place.[11] This means a lot for the Western claim of history's dynamic character, including morality's role. A fertile soil was prepared for modern developments, especially when Joachim's notions turned out to function as a strong, influential motivation for Enlightenment thinkers.

Modern philosophical thought often uses the Trinitarian metaphor in idealistic concepts.[12] Philosophers such as Gotthold Lessing (1729–81) and G. W. F. Hegel (1770–1831) used the Trinitarian scheme to construct their philosophy of history. In their models, we live in the era of the Spirit, an era that features the morality of the free enlightened human being. This interpretation of the Spirit can be considered highly influential for modern economic, political, and religious developments. In the economic sphere, Karl Marx (1818–83) accused Adam Smith of using the Trinitarian formula to hide the real cause of the creation of money. And in the sphere of theology, Karl Barth (1886–1968) built his impressive *Church Dogmatics* on the Trinity of God.

These are just a few examples of the numerous ways in which the Trinitarian symbol was used. Considering the meaning that was attributed to this symbol, its influence on Western thinking and acting is virtually inestimable. It enabled the maintenance of a monotheistic culture and religion, along with the concurrent development of a certain range of freedom. The Trinity allowed for adjusting the figure of God to fulfill a specific purpose or need. We are very familiar with this adjusting: He was in heaven, now he is on earth; he is of and among the people; he is the 'First Cause,' the Architect, the 'Watchmaker.'

In Christianity, the meaning of the word 'God' is connected to the unification of the three 'persons' who symbolize the one true God. The three-person division characterizes God's activities as Creator (God the Father), Reconciler (God the Son), and Sanctifier (God the Holy Spirit). The Trinity expresses the three major 'action fields' that constitute godly perfection and unity. The three-person bond is malleable; the activities can be considered corporeal or transcendent; and simultaneously, there is space for both the revelational character on God's side and the adjustment of meaning on the human side. Thus, Trinity's community symbolizes and guides human community.

Ironically, this subtle symbolic wording is a sophisticated religious system. God the Creator does not show himself through natural reality only. If that were the case, humans would have to submit to the laws of nature. The human species would follow natural instinct and therefore would be considered a mirror of nature. Yet when God shows himself as a human through the Son, humans are elevated to a much higher status: humans represent God on earth. But they can do this only with the help of the third 'person' of the Trinity, the Holy Spirit. The Spirit can be interpreted as the ethical force that connects norms and behavioral rules with convictions, such as the belief in God. The traditional church doctrine proclaims the Spirit's emanation from the Father and the Son ('filioque').

By contemplating this peculiar theological theme, we are able to uncover some typical, dominant aspects of Western thinking. Assuming that these divine characteristics are human projections, they reveal certain aspects of how people think about themselves in reality.[13] Many contemporary scholars consider Christianity and Judaism anthropocentric religions that significantly contributed to the present fatal division between humans and nature. The dominant attitude of the human arises from this division.

In early modern thought, especially Protestant, philosophers and theologians felt free to push God into the distance. Indeed, from the seventeenth century onward, God is localized farther and farther away. This changing image of God is closely related to the mechanization of creation. Originally, creation was considered a living organism wherein humanity had an organic place.[14] But a distance between the 'living' human and the 'dead' creation soon developed. In deism,[15] God is pictured as a watchmaker. The mechanically well-functioning creation reveals his being and his divine will.[16] Imagined as an architect, God does not have much interaction with his own created 'product.' Instead, creation's potential as a 'machine' is revered. This is all we know of a God-creator. Yet it is enough to assure politics, science, and technology of his will and to adapt moral philosophy to natural law and natural right. The origins of these different fields of science and philosophy are located in a religious culture.

Economics could be interpreted as the elemental divine science, helping nature to yield its richness to human beings. For this reason, I term modern politics the Son, conducting the will of the economic Father, thus creating salvation through obedience. The right moral behavior is covered by the Holy Spirit, who emanates from the Father as well as from the Son.[17]

Today, politics often supports the efforts of environmental ethics to make 'naturality' an *idée directrice* (an idea that gives direction) and to overcome the profit principle's ideological power.[18] Yet politicians will never do this without listening to their economic 'boss' with one ear, as the Son conducts the will of the Father. The result is a split attitude, since two lords need to be served at the same time: both the ecological living world and the economic-technical system.[19]

Since *this* God the Father is a tough, unmerciful patriarch, this double loyalty is unsuccessful in everyday practice. In an ambiguous way, he promises that strict adherence to his commands will eventually profit all. Yet in order for the poor to share the wealth and to adequately manage environmental problems, huge amounts of money must be made. To achieve this, a particular Spirit is wanted—a Spirit that convinces through ethical argumentation and ideological standards, one that gets its arguments from conviction: the belief in the Father. The circle is then complete.

Applying this symbolic Trinitarian discussion to individuals, we can assert that modern bourgeois individuals project themselves onto God's throne. Modern bourgeois individuals want to be autonomous, they want to shape their own worldview and ethics, and they have had enough of ecclesiastical and religious paternalism. In the words of philosopher Immanuel Kant, Western, enlightened individuals bear the law of virtues within themselves, as they have the starry sky above them. These humans make their own acting a source of wealth and richness, they carefully tune and prepare their own acting for the economic acting. I am, of course, speaking of the religious, political, and economic elite, whose splendor radiates on an increasing number of people.[20] Whether we can hear God's voice in ethics is a particularly ironic question, but it is to the point. 'God' is the conscience of the well-raised, decent bourgeois. The conscience is then used as a standard for moral acting. In a society where the free-market principle rules, ethics concentrates, in a hidden way, on self-interest. This 'God' can do little for real environmental change.

WHY DEEP ECONOMY?

Both economy and ecology have deep dimensions. Before articulating a clear preference, we must have a thorough analysis—nothing less than a 'collective psychoanalysis'—of both worldviews. The first step is an in-depth analysis of Western 'sources of the self.'[21] The whole heritage should not be dismissed in a few words, which happens in Devall and Sessions's assertion:

> The deep ecology norm of self-realization goes beyond the modern Western *self* which is defined as an isolated ego striving primarily for hedonistic gratification or for a narrow sense of individual salvation in this life or the next. This socially programmed sense of the narrow self or social

self dislocates us, and leaves us prey to whatever fad or fashion is prevalent in our society or social reference group.[22]

These sentences could be the outcome of a thorough study of Western culture's history. Yet lacking this kind of analysis, they are dangerous generalizations. This is why I propose a concentration on the moral, political, and economic thoughts of key thinkers in Western history. Thinkers who are oriented toward ecology and biosphere have to cope with the triad of morals, politics, and economy. The concept of community makes this perfectly clear.

The concept of community plays an important role in the current environmental discussion, as well as in historical reflection on economic philosophy. Daly and Cobb, for example, refer to Aristotle to clarify that 'community' is the social concept that lies at the roots of Western culture.[23] These authors are in search of a new kind of human economics, one that contributes to the well-being of the human community. Hidden behind this plea for an economy that serves the human community and nature, however, is a severe critique on our capitalistic economy. What we need is an economy that truly serves living beings in their relations. We must abandon exploitation, growing capital, and all other community-offending practices.

The lack of reflection on what modern economy really is constitutes the central problem with this approach. In fact, no serious analysis of economy as deep economy can be undertaken as long as economy is looked upon as a monodirected skill, one that will be perfected once its glitches are identified. In this book, I treat the economy in its entirety, including its works and effects. I do not ask merely how the free-market economy works but also why it works the way it does. There are several subdivisions to this question, including epistemological presuppositions, historical and social developments, political and ideological contexts, moral acceptance and apology/defense, and empirical consequences.

The use of argumentation, derived from nature and natural relations, is quite striking in past and present research concerning these areas. The methodology of economic and ecological reasoning proves to be closely related, in spite of the fact that the individual aims and consequences are in opposition. As Pete Gunter says:

> Pragmatism, Marxism, scientific humanism, French positivism, German mechanism: the whole swarm of smug antireligious dogmas emerging in the late eighteenth and nineteenth centuries and by now deeply entrenched in scientific, political, economic, and educational institutions, *really do not, as they claim, make man a part of nature*. If anything they make nature an extension of and mere raw material for man.[24]

We should at least ask why these dogmas insist on incorporating ideology that "make[s] man a part of nature." We must also look at the type of reasoning involved, which uses 'nature' as the ultimate and final argument. So far, organic thinkers and authors have been unwilling to address the relationship between deep ecology and deep economy.

There are several probable reasons for the lack of interest and investigation. One is that U.S. wilderness defenders and biosphere advocates are bored with the European heritage of enlightenment. Another probable reason is that the century-long capitalist-Marxist debate made serious objective reflection on capitalism's nature practically impossible, especially in the United States. However, since the fall of Russian communism and the adaptive state of existing communism, an opportune time for investigation, analysis, and discussion has arisen. Moreover, as the battle between economic superpowers, such as Japan and the United States, intensifies, the need and opportunity for critical analysis of economy's ideological use of nature increase.[25]

There has been a growing awareness of the interconnectedness of humans, other species, and nature in the last decade. Indeed, we have seen a growing consciousness regarding the fundamentals of human community and human enmity, of psychology and ethnicity, and of the religious and moral elements of culture. And more than ever, the questionable role of our free-market system is at the forefront of the ecological and economic arena. Although the issue of exploitation has been routinely addressed, exclusion is now being recognized as a serious problem. In the free-market system, those who do not have the natural richness, the skill, or the technology to participate are banished to an isolated economic death. These issues are of the utmost importance to the concept of community that dominates today's discussions.[26]

The disparity that the economy sustains calls for deep insight into the absolute character of modern economy and its dominating influence on human relations and convictions. At the heart of the situation is the historical development of the human spirit and material relations. It involves our interpretation of life and the world. History shows us that most changes happen in an evolutionary way. Yet looking back, we can see some sudden ruptures—key seismic shifts in Western civilization's interpretations and convictions. These ruptures are important in the current ecological discussion. Scholars and researchers are now studying premodern, and even ahistorical, models of sustenance, community, and economics. The question at hand is whether it is possible and desirable to bridge historical ruptures, and if so, how we should bridge them. Regardless of usage, it is necessary to thoroughly understand the character of these ruptures.

In order to do this, I am invoking major Western thinkers who lived during periods of decisive change for the purpose of close study and analysis. Aristotle, who saw the first wave of Western enlightenment, and Adam Smith, who saw the second, both commented on developments that proved crucial for posterity. Aristotle and Adam Smith have therefore been chosen as our main historical spokespersons. Both are representative examples of 'deep economic' thought within their own contexts.[27]

Once we consider nature and biosphere as ontological entities complete with moral standards, we enter an uncomfortable realm.[28] This is evident in much modern deep ecological thought, as well as in historical treatments of 'man and nature' in general. A risk is involved in the serious examination of the relationship between humans and nature. When ecologists work with biological models, there is a temptation to neglect human history. This is a real hazard. As Daly and Cobb remind us, however, all metaphysical thinking needs correction, because of the 'fallacy of misplaced concreteness.'[29] In fact, searching for obscure sectarian figures is unnecessary; the world's most renowned writers have already informed us of this realm. Our interest now is the rise and development of modern Western economics and its ideological heritage.[30]

Take, for example, Adam Smith's reception of classic concepts and interpretations. If a historical rupture between the classic and the modern approach to social reality exists, we will surely find it here. I highlight some major differences and indicate the decisive turn in Western moral, political, and economic thought. I also outline the role of religious development in the change. After this has been accomplished, we will be in a position to look anew at human and ecological communities in order to determine their potential for saving the oppressed earth and its endangered human inhabitants. I begin by examining the structure and conceptions of Aristotle's ancient Greece.

There are similarities and differences between the old Greek political community, the polis, and modern Western society. Although Aristotle is often characterized as one of Western bourgeois culture's patriarchs, his concept of nature differs considerably from ours. Nature and society (polis) are not divided in a dualistic way. Aristotle considers the human community to be a natural entity. Social reality has a metaphysical structure that is directed toward the final aim of happiness (*eudaimonia*). The function of politics, the art of establishing and conserving the ideal polis, is to dictate ethics, or right acting.[31] It is important to understand that in Aristotelian thought, what is right is 'natural' (*phusei*) and what is wrong is 'unnatural' (*para phusin*). Thus, the polis is a natural-social construction, which subsequently implies a natural-moral construction. In Aristotle's concept of economics (*economia*), three aspects are strongly related: slavery, economics, and politics.[32]

Aristotle identifies the right political community as part of a strict hierarchical construct. The polis represents the highest level of the natural order. The polis-community depends on the maintenance and the sustenance of this order. Thus, in the interest of sustainable development, the human will to change and the desire for wealth need to be tamed.

We can put Aristotle's natural-metaphysical scheme of things into a model of hierarchical interdependent elements. In this model, the higher level controls the lower, and the lower sustains the higher:

1. *eudaimonia*: happiness
2. *polis*: city-state
3. *oikia*: family
4. *douloi*: slaves
5. *barbaroi*: barbarians
6. *thèria*: animals
7. *zooia*: living entities[33]

All elements in this model serve the highest aim; all of them are directed toward achieving the telos of *eudaimonia*. There is a strong connectedness among the individuals of a certain position, and they must cooperate in the realization of the preordained structure. Indeed, this is their divine task as creatures. The polis is the community, which most often approaches the highest good.

In the beginning of *The Politics*, Aristotle uses the following as a connecting passage from Ethics to Politics:

> Observation tells us that every state (*polis*) is a community (*koinoonia*), and that every community is formed with a view to some good purpose. I say 'good,' because in all their actions all men do in fact aim at what they think good. Clearly then, as all communities aim at some good, that community which is the sovereign among them all and embraces all others will aim highest, i.e. at the most sovereign of all goods. This is the community which we call state (*polis*), the community which is 'political' (*koinoonia politike*).[34]

Aristotle's emphasis is on cooperation. In the sphere of politics, free male citizens have to cooperate in order to organize and govern the city-state. These 'officials' must meet certain criteria; for instance, they must be free to perform the political task, which means that their social rank affords them the time to attend to public office. These criteria presuppose a scrupulously defined relation with slaves and with money. These aspects turn out to be determining for Aristotle's conception of the human community as the dominant part of the whole natural reality.

SLAVES

Aristotle makes a clear distinction between the sphere of the polis and the sphere of the *oikos*. On the hierarchical scale, politics is superior to economics. Yet the well-functioning household is an absolute condition for a virtuous city-state government. Aristotle defines the connection between lord and slave as a natural, organic link. This cooperation between lord and slave is comparable to that between the lord and his wife. Both types of relationships are necessary and fruitful, and they are directed toward particular goals. The lord creates new life with his wife and life-sustenance with his slave. Thus, the *oikia* is the condition for good government of the polis, because it frees the lord. The free male citizen is the lord of the house (*oikodespotes*), but his life is 'reproduced' in several ways.

In Aristotle's view, the social order is the natural order. Every level is part of a totality, and the position is fixed in this metaphysical hierarchy. A fundamental change is forbidden because *this* is the order that guarantees optimum *eudaimonia*. The lord's wife should function in a virtuous way as spouse and mother. Public office is not within her reach. Slaves should function in the right way; one could call this the slave's virtue. The slave cannot become a free citizen. Barbarians can become slaves, although they cannot rise any higher. Slave and wife are the core of the *oikia*; the well-being of the broader community depends on their functioning. The lord's task is to provide virtuous leadership.

In discussing the government and organization of the polis, Aristotle does not mention the slave. Although the slave is the pivot on which the sphere of *economia* depends, he or she simply does not exist at the political level. The slave is not considered human in the free, democratic sense of the word. In Aristotelian thought, this social order is in accordance with nature (*phusei*). This is a point to remember every time people advocate a polis-like community.[35] In order to understand the slave's position, we need to understand the role of money in Aristotle's political and moral thought.

MONEY

The monetary relationship is judged from the standpoint of the polis-community's well-being. Aristotle distinguishes between natural and unnatural economic means: slavery is natural, whereas money is unnatural. Of course, this is the exact opposite of modern economics.

1. very natural: knowledge of the household (*oikonomike*)

2. natural: acquisition of property (*ktetike*)

3. natural/unnatural: exchange, trade (*allage, kapeleia*)

4. unnatural: creation of money (*chrematistike*)

5. very unnatural: rent/loans for profit (*tokismos*)[36]

In Aristotelian society, the natural should never be overruled by the unnatural. 'Natural' relations are stable, and they support the entire construct of social reality. Acquisition of property is necessary for the common good: barbarians captured through warfare are considered property, and the owner undertakes their training as well-functioning slaves. Exchange and trade are natural distribution means. When one has too much of commodity X and is in need of commodity Y, direct exchange is an obvious, natural act. Karl Marx incorporated this into a scheme for modern economics: C-M-C^1 (commodity-money-commodity 1). The system remains stable only when money is used exclusively for the trade of utilities. However, once money is used as a general means of exchange, an unnatural element is introduced. Aristotle condemns moneymaking and the accumulation of capital as very unnatural. Marx charts this as M-C-M^1. In this case, use value is converted to exchange value. Aristotle warns that this trade reversal would jeopardize the human/natural community (scheme 4, above). Indeed, the creation of money is unnatural, and interest, based on its ability to give birth to new money, is very unnatural (*para phusin*) (scheme 5, above). Once the potential of money is realized, there will be no foreseeable end. When the making of money dominates *economia*, the polis-community will perish.

The power and terrible potential of money and capital were foreseen as early as the fourth century B.C. Aristotle's condemnation of *chrematistics* arises from deep ontological insight. It is important to remember that oppression based on social equality was not part of Aristotle's cultural perspective.[37] On the contrary, the metaphysical order of reality dictates that social inequalities will eventually guarantee all happiness. This is a classic 'deep' economic notion. It is 'deep' because the polis's economic organization is deeply rooted in the Greek worldview (presented in above scheme). In Aristotelian society, politics is a morally responsible practice that sustains this metaphysical structure of reality, whereas *economia* relates to the organic structure of reality and is subordinate to politics. Man is defined as *zoon politicon*: the male, white, free civilian; his position is made possible by women and slaves. This kind of 'connectedness' is the presupposition of Aristotle's strong resistance to *chrematistics*. It is a moral-metaphysical judgment.

MODERN ECONOMY AND THE FREE INDIVIDUAL

Aristotle's main question concerning deep economy is the relation between the different elements that construct reality. Aristotle is a 'deep' economic thinker because in his conception, economics concerns the fundamentals of the community; economic rules and practical economic acting serve the social-political construction of reality. The primacy of polis over *oikos* is very clear.[38] In depth, economics is judged by morals, and morals aquire their standard in the metaphysics of reality, termed 'nature.' Within this scheme, free moneymaking, by societies or individuals alike, is highly unnatural and thus bad. All kinds of capital accumulation are considered unnatural. What a society needs is free people with mediocre property (*ktesis mese*), not very rich or very poor individuals. As much as possible, the structure of society should inhibit the attainment of large wealth.[39]

Adam Smith, living some twenty-one centuries later, on top of enlightenment's second wave, confronted a major problem. He, too, wanted to base economics in morality, yet there were fundamental changes in society after the Middle Ages.[40] The ancient *economia* had been undermined, because labor was no longer available in 'natural' slavery. Although slavery was not yet wholly abandoned, it was no longer considered a 'natural' phenomenon. The idea and ideal of freedom and equality for all individuals became the ruling ideology.[41] Due to these changes, a radical new concept was necessary for modern economics. This concept had to reach the radix, the roots of society. Thus, like Aristotle, Smith was concerned with deep economy. Smith's task was to determine what type of human community and connectedness would allow for the creation of an economic conception that guaranteed the 'wealth of the nations.' Smith's moral and economic arguments and the ideological character of his reasoning had to bridge the aforementioned historical rupture. Interestingly, like Aristotle, Smith grounds his theory in the conviction that 'natural' relations construct reality.

One of the significant changes from ancient to modern Western social theory is the value reversal in estimating *chrematistics*. What Aristotle unconditionally condemns as unnatural and immoral, Smith welcomes as a very natural possibility. A similar reversal occurs in estimating the 'natural' condition of slavery. Central aspects of Smith's thought and methods are outlined here:

1. In describing the modern economy's (free) market character, Smith concentrates solely on exchange and leaves the process of production in the dark, particularly the issue of money's creation.

2. Smith writes his moral philosophy on virtues and the virtuous life together (theory of moral sentiments) as a theoretical moral foundation of modern economic society.

3. Smith's concept of 'nature' and 'natural' is built on a deistic kind of natural theology.

4. Smith ultimately fails to explain money's origin and creation, and he

neglects to address the exploitation and exclusion of human individuals and nature.

In order to understand Smith, we must recall his eighteenth-century European context.[42] The Industrial Revolution made exchange possible on a great scale, one that was unimaginable to ancient and medieval cultures. This complex development was accompanied by an outburst of knowledge and autonomy in the fields of science, technology, politics, and economics. Culture and nature were threatened in diverse ways as old opinions, values, and relations decreased, as age-old wisdom and skills were overrun, and as tradition was forced to give way to progress. The new economic order's interest in nature was based entirely on profit. In the interest of profit, old community structures were reorganized for production, exchange, and consumption. A new connection between labor and property developed, and economic and political power were redefined. Ultimately, these new European ideas and practices conquered the world.

In this complex process, economy played an important role. It concerned technical and organizational problems as well as moral, psychological, and religious aspects. Real progress and development are possible only when people, especially people in power, are convinced that well-being and wealth are the final aim. Just like Aristotle, modern philosophers are concerned with the question of how society's structures protect the common good. In doing so, they help legitimize the choices that are made. Descartes' dualism, Kant's individualism, Hegel's concept of history and spirit—all of them are deeply engaged in a giant legitimizing process. They all reflect on relations: human communities and the relations between humans and other species. In the nineteenth century, Charles Darwin appears as a champion of the Western approach to nature.

Before turning to Adam Smith as the most important thinker regarding the place of economics in the entire context of progress, we should pay some attention to his fellow countrymen Thomas Hobbes (1588–1679) and John Locke (1632–1704). Neither of them developed a coherent economic theory, but both had some fundamental ideas on building a new, commercial type of society in the old European Christian culture. I concentrate for a moment on both thinkers' theory of money as a core element of early modern society.

In the introduction to his edition of *Leviathan*, Michael Oakeshott calls the book a myth: "In it we are made aware at a glance of the fixed and simple centre of a universe of complex and changing relationships."[43] Hobbes introduces the commonwealth as an organic system of power, a "mortal god, to which we owe under the immortal God, our peace and defence."[44] After lengthy political-theological considerations, Hobbes turns to the Leviathan's economic character, under the heading "Of the Nutrition, and Procreation of a Commonwealth" (chapter 24). Here, Hobbes claims:

> Money, of what matter soever coined by the sovereign of a common-wealth, is a sufficient measure of the value of all things else, between the subjects of that commonwealth. By the means of which measure, all com-modities, movable and immovable, are made to accompany a man to all places of his resort, within and without the place of his ordinary resi-dence; and the same passeth from man to man, within the commonwealth; and goes round about, nourishing, as it passeth, every part thereof; in so much as this concoction, is as it were the sanguification of the common-wealth: for natural blood is in like manner made of the fruits of the earth; and circulating, nourisheth by the way every member of the body of man.[45]

For centuries, this foundational statement has served as a basis for develop-ing theories of money. Here, Hobbes shows a stark intuition concerning money's circulation as the presupposition for a modern, dynamic society. But he also claims the beginning of an idea of money's creation: "for natu-ral blood is in like manner made of the fruits of the earth." In retrospect, we can presume a first sign of consciousness of the process of production and exchange to come: the process of making money "of the fruits of the earth." But this being the seventeenth century, we should not overanalyze this theory. Once and for all, however, money's absolute indispensability is made perfectly clear.

John Locke's *An Essay on the Value of Money* considers the question "whether the price of the hire of money can be regulated by law" and an-swers right away: "it is manifest it cannot."[46] According to Locke, money is a means to get wealthy, and for England, money is earned by trade: "For, money being an universal commodity, and as necessary to trade as food is to life, everybody must have it, at what rate they can get it."[47] Locke vehe-mently advocates a free market, both for money and for other commodities. Hindering trade is life-threatening:

> For, we having no mines, nor any other way of *getting*, or *keeping* of riches amongst us, but by trade ... the overbalancing of trade between us and our neighbours, must inevitably carry away our money; and quickly leave us poor, and exposed. Gold and silver, *though they serve for few, yet they command all the conveniences of life*, and therefore in a plenty of them consists riches.... Trade, then, is necessary to the producing of riches, and money necessary to the carrying on of trade....
>
> Money, in its circulation, driving the several wheels of trade, ... is all shared between *landholder*, whose land affords the materials; the *labourer*, who works them; the *broker i.e.* the *merchant* and *shopkeeper*, who distributes them to those who *want* them; and the *consumer*, who spends them.[48]

In Locke's remarks on money, we encounter some of the same ideas as in Hobbes. Locke, however, took some big steps on the road to a free-market society, bringing together private property, trade, and money in a well-functioning system as a moneymaking machine. Money itself is the magic component that links (Locke speaks of the *sharing* of money) the different parties that keep the machine running.

In Adam Smith, Hobbes and Locke come together and are lifted to a higher level. Standing on his predecessors' shoulders, Smith can be considered one of the foundational theoretical defenders of the system in the making. The possibilities of the free market fascinated him. He saw exchange as a many-sided blessing: people have commodities and utilities at their disposal, production is stimulated (division of labor), and money grows as a means of exchange. Smith celebrates the free and equal subject. All (male) individuals have the same chance; they are free and faithful partners.

Human goodness and faithfulness were not self-evident assumptions in eighteenth-century Britain. Hobbes had claimed the very opposite a century earlier. Hobbes upheld the sovereign as the divinely appointed one with the task of forcing individuals in the right direction. Here we find the core of Smith's 'enlightened' vision. The divine no longer resides in heaven and appoints an all-powerful earthly representative. On the contrary, now the judging power is incarnated in every well-educated individual. In Smith's words, it is "the demigod within the breast."[49]

Here we meet the intellectual spirit of the Enlightenment. Education of the citizen will bring forth a community of equals who are aware of their interdependent positions. Being members of a community, they will focus not only on their own interests but also on others' needs. As Smith claimed, 'man without' corresponds to 'man within.' Of course, self-interest is the point of departure, but all members know that they depend on others for the fulfillment of their own interests. They also know that money is the only condition by which one accepts others and is accepted by others. As Smith states:

> In order to avoid the inconvenience of such situations, every prudent man in every period of society, after the first establishment of the division of labor, must naturally have endeavored to manage his affairs in such a manner, as to have at all times by him, besides the peculiar produce of his own industry, a certain quantity of some one commodity or other, such as he imagined few people would be likely to refuse in exchange for the produce of their industry.[50]

In addition to having money "at all times by him," Smith advocates an education in justice, prudence, benevolence, sympathy, and other virtues. A community that respects the common good, practices commerce, and circulates money requires virtuous members. Note that the word 'common' is the base of 'community,' 'common good,' and 'commerce.' The concept of 'common' is the main difference between the premodern and the modern community. For the elaboration of deep economy, this is the core element. In contrast to Aristotle, Smith concentrates community in the individual, who behaves in a manner that all other individuals accept. Everything depends on the individual's imagination.[51] Eventually, the individual's behavior will be in accordance with the social-political-economic system; cooperation is achieved through the individual's will and imagination.

Sympathy exemplifies this cooperative system.[52] This virtue is fundamental to the functioning and well-being of the community. First, individuals must consider their own conduct. For example, I have to look at myself from your perspective in order to gauge how you will react to me. I try to read your countenance and body language and determine your criteria for judgment. I judge my own behavior as if I were you. In order to do this, I must be able to transcend the boundary of myself. Doing so demonstrates that I possess transcendent power. Adam Smith terms this the "impartial spectator."[53]

According to Smith, sympathy is a mediating concept between benevolence and prudence. Benevolence is the divine virtue of altruism. Although humans are social beings, totally dependent on one another and the community system, they are unable to live in an orderly and harmonious way. Human nature is inclined toward self-interest. In order to survive, one must exercise prudence, which is well-understood self-interest. This famous passage from *The Wealth of Nations* defines the content of prudence:

> In civilized society he stands at all times in need of the cooperation and assistance of great multitudes, while his whole life is scarce sufficient to gain the friendship of a few persons. In almost every other race of animals, each individual when it is grown up to maturity, is entirely independent, and in its natural state has occasion for the assistance of no other living creature. But man has almost constant occasion for the help of his brethren, and it is in vain for him to expect it from their benevolence only. He will be more likely to prevail if he can interest their self-love in his favor, and show them that it is for their own advantage to do for him what he requires of them. Whoever offers to another a bargain of any kind, proposes to do this. Give me that what I want, and you shall have this what you want, is the meaning of every such offer; and it is in this manner that we obtain from one another the far greater part of those good offices we stand in need of. It is not from the benevolence of the butcher, the brewer, or the baker, that we expect our dinner, but from their regard to their own interest. We address ourselves, not to their humanity but to their self-love, and never talk to them of our own necessities but of their advantages. Nobody but a beggar chooses to depend chiefly upon the benevolence of his fellow-citizens.[54]

Civilized, economic society is based on a principle that is different from altruism or egoism. It is based, in fact, on the self-love and advantage of the other. Sympathy is the agent that harmoniously mediates between my needs, desires, and necessities and the other's self-love and self-interest. This concept is rooted in the Stoic tradition. The Stoic idea of a cosmic *sym-patheia* that commands and covers the entire universe, including human community, underlies Smith's preference for the term. The opening of Smith's *Theory of Moral Sentiments* tells us that humans are not just egoists:

> How selfish soever man may be supposed, there are evidently some principles in his nature, which interest him in the fortune of others, and render their happiness necessary to him, though he derives nothing from it except the pleasure of seeing it.[55]

In this context, sympathy is an attitude that functions in three phases:

A (the self) sympathizes with the feelings of B (the other): (A → B)

B exercises a type of self-control that enables sympathy with A: (A ← B)

AB: In this reciprocal relationship, both parties notice and sympathize with each other (A ↔ B). Each attempts to identify with the other, even though they both know that complete identification is impossible.[56]

As Smith states:

> And hence it is, that to feel much for others and little for ourselves, that to restrain our selfish, and to indulge our benevolent affections, constitutes the perfection of human nature; and can alone produce among mankind that harmony of sentiments and passions in which consists their whole grace and propriety. As to love our neighbor as we love ourselves is the great law of Christianity, so it is the great percept of nature to love ourselves only as we love our neighbor, or what comes to the same thing, as our neighbor is capable of loving us.[57]

Smith describes modern society's constitution and simultaneously summarizes the entire Western tradition:

1. A fusion between the Christian commandment of love and the Stoic morality of self-control.
2. A combination of the non-Christian antique tradition and the Judeo-Christian tradition.
3. The synthesis of Christian revelation and the natural law.

Smith's interpretation of the Great Commandment (Matthew 22:39 and Matthew 7:12) is such that it becomes a prescription for self-love without the abandonment of love for the neighbor. Thus, self-love and neighborly love lose their extreme character, thereby becoming manageable virtues. Notice that the Second Commandment changes from a unilateral command of love to one that incorporates a reciprocal relationship of give-and-take.[58]

The important condition for the execution of this commandment is that one must learn to look at oneself through the eyes of another person. This means that through confrontation with myself, I become a different person. Smith calls this judging principle the impartial spectator. (We could indicate it as C, circulating the A ↔ B relationship.) There seems to be a judging authority other than traditional opinion, which is no longer sufficient. By what or whose authority does this impartial spectator principle operate? Neither God as a heavenly power nor the metaphysical concept of 'nature,' as in Aristotle's hierarchical order, is an acceptable operating principle now. Smith localizes this human and divine authority in the individual. The impartial spectator has a "demigod within the breast." In theological terms, God is incarnated in the modern Western individual.

Even if we accept this precept, we are left with the question of what or who is this superego? Is it society, tradition, or just a natural possibility in human beings? Smith gives physical, mathematical, philosophical, and theological reasons and arguments to achieve ambivalence on this point. Smith's theory rests on his conviction that human moral consciousness is a natural force, and its operation is analogous to the laws of gravity.

The individual trained in virtue will be able to control his egoistic inclinations. But some type of Creator or Divine Nature is present in nature. In Smith's deistic thinking, God and nature are identified within the context of human responsibility and conscience. God is a virtual being. God is the person's own mirrored image. Smith combines this concept of nature/God as he defines it in the *Theory of Moral Sentiments* with the human trading community described in *The Wealth of Nations*.

The moral function of egoism is superior to its devastating, immoral function.[59] Under certain conditions, egoism can be considered a moral engine. One of these conditions is justice; another is fairness. This is the structure of community. This is where economics needs to provide a moral, philosophical, and theological service and meaning to the human community that uses it. Smith's documentation of economics' positive character indicates his awareness of this need, yet his silence about economics' dark side reveals his own blind spot.

ECONOMY'S VIRTUOUS CHARACTER

We have collected the basic criteria needed to understand the relationships that constitute deep economics. We can articulate it as follows:

1. Egoism's driving force is tamed by the mediator, sympathy.

2. The abstract human expresses a part-divine, part-human nature through the combination of benevolence and prudence, a combination that is achieved through sympathy.

3. The judgment and authority of the impartial spectator come from within the individual, the abstract person.

These criteria are crucial to the ideological structure of modern economy, because the modern concept of community radically differs from the ancient idea of the polis. The ancient political society changed and became the modern commercial society. According to Smith, human community is no longer a metaphysical, hierarchical construction. Individuals communicate by exchange. The virtues are the basis of commerce. Individuals are acceptable to others when they succeed in making themselves acceptable. In the moral sphere, the imagination of the other's feelings directs one's own behavior. Free, well-educated individuals have the natural capacity to form a moral community. They realize this possibility under the condition of moral self-control. In the economic sphere, individuals are acceptable to others through exchange. Individuals must have money to be accepted in this sphere.

By incorporating the imagination, Smith combines moral and commercial possibilities in a subtle way. From a comparison of *Theory of Moral Sentiments*[60] and *Wealth of Nations*,[61] we can conclude that the impartial spectator is transformed from a human-divine category into "some one commodity or other, such as he imagined few people would be likely to refuse." Thus, to be an acceptable member of a (commercial) community means that one is virtuous and that one commands money. Both aspects, the moral and the economic, are interchangeable. The interchangeability of commercial society and moral community is such that moral conditions facilitate the working of the 'invisible hand.' Indeed, they make this working possible. The reason for this is their mutual natural character. Smith asserts his moral theory on theological foundations. Human moral nature has divine characteristics: God reveals himself in the modern individual. This is modern, deistic Christology. In economic theory, the wealth of the nation depends on the right commercial attitude. This commercial attitude cannot be reached without the right moral attitude.

Smith's position entails a fundamental denial of the historical antagonistic structure of economically defined social classes. Of course, Smith acknowledges that there are rich and poor people. But he asserts that these differences will decrease eventually, if not vanish altogether. Everyone will be able to survive in the communities created by commercial society's development. Smith presupposes the presence of money. In his abstract view, everybody can get it and participate in the commercial society. Everybody has access to the exchange market. According to Smith, there are enough financial sources and resources: "Wages, profit and rent are the three original sources of all revenue as well as of all exchangeable value. All other revenue is ultimately derived from some one or other of these."[62]

This is what Marx, commenting on Smith's theory of money, called the 'economic Trinity' or the 'Trinitarian formula,' the core definition of modern economic philosophy. Smith assumes that all members of the commercial society are able to participate in one way or another. The rich will not be able to consume all they get, and the poor will receive from the abundance of the rich. The presupposition of the growth of wealth and of progress is characteristic of the commercial society. The 'invisible hand' distributes the blessings.[63] It is the nature of money to duplicate itself. Once one succeeds in gaining a small amount, there is no limit. Smith claims that the starting phase of the individual wealth process is the most difficult one. It takes only a small amount to begin, but "the great difficulty is to get that little."[64] Smith uses Holland as an example of great success, "where almost every man should be a man of business or engage in some sort of trade."[65] The determining variable is one's own moral behavior. Industrious people will certainly succeed.

Smith foreshadows one of the most striking characteristics of Western ideology. Reasoning starts at a point where people feel comfortable. The foundations of the commercial society can be found in the human community. Money, the magic means of wealth, simply exists in the form of wages, profit, and rent because people developed a system in which property and labor are the money-creating forces. Smith never asks *whose* property and *whose* labor. Terms are abstractly interchangeable: laborer, owner of means of production, and landlord often seem to be the same person—a 'Trinitarian laborer,' perhaps. The famous self-supporting gardener makes this clear:

> A gardener who cultivates his own garden with his own hands, unites in his own person the three different characters, of landlord, farmer, and laborer. His produce, therefore, should pay him the rent of the first, the profit of the second, and the wages of the third. The whole, however, is commonly considered as the earning of his labor. Both, rent and profit are, in this case, confounded with wages.[66]

The gardener symbolizes people cooperating in the moneymaking process. Smith speaks of three characters that can be localized in one person as a sort of corporate personality. The individual alone forms a community. When persons or characters use their virtuous human capacities, wealth is guaranteed. This is part of deep economy, in that it is the combination of morals, connectedness, and economic knowledge. Once the blessing of *chrematistics* is recognized and accepted, the majority no longer considers moneymaking hazardous. The 'deep' character is the quality of natural legitimacy. Smith compares the laws of the social system with the laws of natural science. The laws of exchange are as consistent and true as the laws of gravity.

Smith's 'deep' theory is typical, in that it does not reach far enough. The theory's flaw is its naïveté toward money. Marx's most important contribution to economic thought is his questioning of the origins of money. Most economists are uncomfortable with Marx's questioning. They prefer to deny the problem of money's original creation. In Smith's theory, there is a hidden assumption that undermines the seemingly virtuous combining process. Something must precede the point where Smith starts. Money cannot be born as wages, rent, and profit. The transforming of labor and nature into money (wages, rent, profit) is itself a process. The examination, description, and questioning of this process must go further than Smith's combining of morals, nature, and exchange. It is not enough to discuss the existence and availability of money, its processes of exchange and consumption. The discussion must reach the nether regions of production where conception takes place. In Aristotle's vision, lord and slave cooperate in a harmonious *economia*. Smith suggests the same ideal cooperation in modern society. The great step forward is the division of labor.[67]

In modern society, however, owner and laborer (employer and employee) are divided. In his effort to go further than Smith, Marx defines both the owner and the laborer as dependent on money, with the distinction that the owner earns money as profit, the worker as wages. The employee sells his labor, producing surplus value. The surplus value is the employer's profit. This profit, realized in the marketplace, is produced in the factory.[68] Marx's analysis undermines all harmonious economic theories, regardless of their 'deep' intentions. The Trinitarian formula does not cover the ultimate, co-operative production of money. Van Leeuwen's unitarian formula more aptly indicates exploitation as the inevitable source of money and wealth.[69] Regardless of whether or not one agrees with this analysis, it casts doubt on the actuality of harmonious living in the Western Hemisphere.

Although I have been discussing structural-strategic positions, working as cornerstones of our moneymaking system, we can also speak in terms of power relations. Gaining money, and thus participating in progress, accumulation, and growth, is possible only when one controls the forces of production. This does not mean the possession of capital only; it also assumes free access to nature, in the form of raw material and energy and as a dumping site. This presupposition illustrates the dichotomy between humans and the rest of nature.[70]

Money can be created only in a process of transformation.[71] Investing money for profit presupposes an action on the original amount of money.

After the second Industrial Revolution, it is possible to 'imagine' that the origin of exchange, trading, and moneymaking lies in communication technology. Yet in the area of 'deep' economy we must realize that our system depends on production, market, distribution, consumption, waste, and so forth.

In this respect, nothing has really changed since Adam Smith, except that the situation has intensified. The financial centers today are banks and stock markets. Their influence on behavior, convictions, and self-control is enormous. This is true on both the individual and the group level. The central point for the deep economist is that the financial centers of power derive their strength from something that precedes, feeds, and maintains their position. The industrial society provides the material means. So in rich countries, state banks usually command huge amounts of money, making this money available to commercial banks at very low interest rates (often 3 percent).

Everyone depends on the banks' general power; no exception is made for deep ecologists or wilderness defenders. It is impossible to be free from capitalist burdens in our society. One can resist, but the omnipresence of the free-market economy is undeniable and ultimately unavoidable. This is the reason that I awarded it a divine character.[72] As Adam Smith describes, we must all adopt virtuous ways in order to survive and support our modern social system. The issue is whether people are aware of the connection between virtues and free-market economics.

MAKING MONEY: COOPERATION OR EXPLOITATION?

In Marx's unitarian formula, there is one way to make money—exploitation. Marx speaks mainly of the exploitation of the proletariat class by the bourgeois owners of capital and production. Today, we must complement Marx's formula, not replace it in any way, with the exploitation of nature. Because of the deep structure of our type of economy (*chrematistics*), we must never focus on the exploitation of people without taking into account the depletion of nature. This is the critical point in environmental discourse. The broad U.S. discussion exemplifies this problem. Daly and Cobb's *For the Common Good* is an effort to transcend the strange lack of interest in economics at a meaningful level. It also indicates that there is a struggle within the field of ecology to deal directly with the critical issues.

The need for money is so great that most humans are willing to exploit nature. I use the term 'exploitation' in the qualified sense of extracting energy and power from someone or something to such an extent that the object from which energy is extracted is unable to restore its original balance. The balance between giving and receiving is lost: the earth gives material and energy but receives only polluting waste. This development parallels the exploitation of labor, particularly the use of labor in low-wage countries. Recently, exploiting activities have been 'reallocated' to poor countries. Cheap labor is not available within the legal circuits of Western societies, where unions are successfully established as part of a democratic political system. The illegal circuits within Western societies and poor, non-Western countries are then tapped for cheap labor. A similar situation has developed on the environmental front: exploitation occurs illegally in Western countries and is widespread in poorer countries, where legislation is generally weak. With the exportation of exploitation to poorer countries, the rich Western countries isolate themselves in a cloud of false global well-being.

There is a tension between cooperation and exploitation. What I call exploitation, another person might call cooperation. Adam Smith thinks that economics concerns right and fair cooperation. Smith is surely not outdated. Most economists, politicians, and other reasonable people express a deep conviction that right cooperation among individuals, groups, and nations underlies economics. Right cooperation includes fair distribution. There is even a distribution of tasks in our system: economics focuses on cooperation, and politics focuses on distribution. Thus, our system requires perfect cooperation between economics and politics.

It is therefore fitting to speak of a 'deep' conviction in this context, but we must ask, isn't 'shallow' a better qualification? The 'Smithsonian' sphere, which connects morals to exchange relations, tries to connect people in a 'deep' way by providing commercial society with moral 'depth.' Moral control is necessary to prevent and prohibit wrong behavior and unfair consequences. The current discussion about sustainable development is a good example. There must be an undisturbed balance between cooperating partners. This, however, should not disturb the economy's continuity, which presupposes a minimum amount of growth.[73]

It is obvious that this type of 'deep' economy does not go far enough. As discussed in relation to Adam Smith, it focuses on the aspect of exchange, and this is an arbitrary starting point. We do not live in a (free) market economy alone. Exploitation precedes cooperation. The market economy can exist in continuity only under the condition of exploitation. The character of money as a means of capital accumulation makes this exploitation necessary. To change this, we have to challenge the meaning and working of money itself.

In the second edition of *For the Common Good* (1994), Daly and Cobb add an appendix on money.[74] They were asked for their opinion on this central issue because they did not cover it explicitly in the first edition. I cover this point extensively in chapter 2, but I mention it now because of its symptomatic character. Most ecological and many economic studies neglect money as the crucial denominator, but as Smith already realized, the possibility of communicating depends on it. The reason for this neglect might be the complicated, multiple meaning of money. Money is like a divine power: omnipresent but only partially and temporally visible. Like the godhead for *homo religiosus*, money is a social power for *homo economicus*. This is so self-evident, that it doesn't seem necessary to question it in a radical, fundamental (deep) way. It simply works: those who control money are entitled to control reality.

The parallel between divine and financial power is not just an illustrative analogy. In modern society, money has a religious character, and it should be analyzed accordingly. The criticism of religion, which began in the eighteenth century, ultimately resulted in the dissolution of its unmitigated power. The economy has been taken at face value long enough; it is time for deep questioning and thorough criticism. Revelation versus religion, clarity or obscurity: this is secular society's dialectical-theological option. We need to establish light in the darkness of 'capital's night.'[75]

We hold the general conviction that our society's well-being depends on the economy's continuity, on 'sound' economic development. Hobbes states that money should be considered the blood of society's body. It is the hidden but vital circulating element.[76] The body parts that blood does not reach are destined to perish. We must remember that although the blood metaphor aptly describes the working of money, humans attribute the meaning, and in this way it differs from physical blood. Reflecting the blood metaphor itself, money is a dichotomy—it is both abstract and concrete. Money functions like blood because humans consider society to be a body. This organic presupposition requires a fluid element to feed its connected organs and limbs. One need only think of the commonality of terminology related to corporate cooperation and organic wholeness to verify the aptness of the blood metaphor, not to mention the strength of people's conception of the social body.[77]

All economic activity starts from the basic fact that humans care for their own lives. In contrast to all other animals, humans, being reason-gifted animals, do so in a rational way. Science and technology are among the ways in which modern humans care for their lives. In order to develop, however, modern humans must share. In the ancient Greek polis, people had to share too, but sharing was practiced under sharp, clear conditions. Certain people were considered humans, others were not. The polis considered money a hazardous element because of its capability to accumulate power. If money acquired power, it would divide and destabilize the social system in such a way that it would be impossible to maintain the established method of sharing.

There has been a radical, decisive rupture in modern times. We consider the ancient Greeks to have been mistaken on two points: We believe that everyone is human and that all are principally free and equal. Importantly, equal humans are able to share and cooperate by monetary means. In Europe today, money has a divine, mediating function; hence its meaning has been

broadened. The historical meaning of money as a means of exchange remains, but money has grown in status as it has increasingly become a measure of value.[78] Stability is necessary to ensure continuity in money's functions.
Modern economists believed that this could be achieved by reviving and
reinterpreting the treasure principle, the accumulation of capital through money.
This economic action, based on private property and private responsibility,
seemed to be the only way to ensure continuous economic development.
Thus, the ancient forbidden act of *chrematistics* is the modern means of
economic growth and social stability. In order to protect money's two other
functions—as the measurer of value and the means of growth and accumulation—money has to be continuously created in an upward spiral. In short,
capital is movement. It is the circular movement of production, exchange,
and consumption within a complicated system of variable components. The
end result, however, must be profit. This end is a fixed absolute.

WINNERS AND LOSERS: THE BLESSED AND THE DAMNED

Economy's original act, repeated day in and day out, year after year, is the
creation of profit. We refer to this profit as surplus value. 'Surplus' means
'more'; the act is based on the confidence that money earned will exceed
money invested. This is the deep meaning of credit. In this system, the
central credo is that the political and moral conditions will ensure safe
investments and rents. One is assured of getting more than one spends.[79] A
condition for this credit-confidence is the freedom to use, to exploit, and to
exclude. This is the main reason that environmental and humane measures
are always limited when economic interests are at stake. Once there is a
demand for utilities and services, there should be no limitations on production,
trade, and distribution. Accumulation, thus stimulated, brings wealth. This
wealth arouses the purchase power that moves the spiral in an upward direction.
   Money's third and most important function is the accumulation of capital.
Money has a magic-like ability to double itself: Money gives birth to money.
How can we account for this magic? According to Smith's line of thought,
once invented, money increases by acts of trade and exchange. Banks and
states create money, and like all fertile creatures, it reproduces itself.[80] The
procreative power of money inspires religious devotion: human trust and
obedience. This cooperation strengthens its power. Underlying this devotion
to capital accumulation, which is believed to be good for the individual and
the community, is the effort to produce and sell goods cheaply within a
competitive framework. This is accomplished by exploiting humans and nature.
It is the original model for modern market relations: The only way to earn
money through the process of buying and selling is to produce as much as
possible as cheaply as possible.

One major presupposition in this process is the concrete act of exclusion. Deep economy's basic model is a combination of cooperation and exploitation, and this presumes exclusion. As in the ecological sphere, the economic sphere excludes the components that have lost the ability to survive within the system. The components that are unable to reproduce and circulate the system's blood (money) are left out. In some instances, these components were once profitable, functioning members, but for some reason, they lost their functional quality. The area of 'for some reason' is where we should focus the analysis of oppressed people and nature. Exclusion can occur both before and after exploitation, and there are deep economic reasons for both cases. In some instances, societies and habitats are of no commercial use from the start; in other instances, they are used to depletion and then discarded.

The interrelatedness of the economy includes component exclusion. The cold reality is that those who do not succeed are considered losers. Regardless of their past status, a current inability to command the means of achieving success excludes one from further participation. The loser is out of the game. Moreover, as long as money is the decisive measurer of value, exchange value as the one dominant meaning of value will remain fixed. The economic system requires that cooperation be between only the chosen, the blessed, the 'saved by saving money.'

Under the iron-clad laws of deep economy, there are three cooperative functions of money. Every day, everyone everywhere uses money as a means of exchange. By doing so, people contribute to the accumulation of capital. Spending money contributes to growth and accumulation. Spending money is the means of exchange, which entails that the spender accepts the seller's conditions. The purchaser trusts that the seller will be able to order new utilities. These utilities must be produced. In order to do this, raw material, energy, and labor work together. This process is essential for the development of a country or region. Banks can finance projects only if the basic conditions of capital accumulation are fulfilled at the core level of the economic action.[81]

Nothing has changed fundamentally since Smith wrote *The Wealth of Nations*. In 'commercial society,' success is guaranteed, and eventually, the wealth of all nations will be established on earth. But as Smith already admits, private property is a condition for success. The failure of state capitalism confirmed this 'universal law.' It is not the accumulation of wealth in general, but rather the particular growth of the individual's wealth, that drives the common wealth's engine. For this reason, there is a steadily widening gap between the very rich and the very poor, as evidenced in the United States.[82] According to general belief, personal interest keeps economic activities alive. Smith's moral philosophy is directed toward mitigating rough, cowboy-like economic relations. His great commandment is partially a plea: Please don't exploit and kill one another, but cooperate and share instead. This can stand as modern society's ethical challenge, but it is also an effort to mitigate the unbearable truth of modern economics.

The issue is not merely how to conquer human selfishness. If this were the point, deep ecological consciousness would probably convert those addicted to the collective desire.[83] Of course, we can consider self-interest as a driving force, the subjective character behind an objective necessity. This 'subjective character,' however, has developed into an institutionalized driving force. It has become an objective quality and, in the gestalt of sympathy, a major virtue. In order to overcome this force, we need to clarify the dichotomy of the human. The human is an organic part of economy who simultaneously sustains economics. In order to begin the process of change, we have to accept the dualistic structure of the modern human.[84]

A European who spends enough time in the United States is sure to be struck by Americans' pride in their wealth. Americans like to show their wealth, and in the eyes of Europeans, they are exhibitionists. Europeans tend to hide their wealth, because deep down, they are ashamed of it. A Presbyterian church I once visited in the United States featured a sermon on the connection between wealth and religious blessing. In a naive and uncritical way, the minister repeated Max Weber's thesis literally. He presented wealth as blessing and poverty as curse. As presumed wealthy churchgoers, we were urged to share our wealth (blessings) with the church.

It is doubtful that Weber intended his theory to be incorporated into religious sermons for self-sufficient, upper-middle-class white Americans. Surely, he acknowledged salvation as an important concept in both the religious and the social spheres. Weber upholds that some are saved and some are damned; some receive blessings, others curses; some are winners, and others are losers. In his famous study on Protestantism and capitalism, Weber describes how seventeenth-century Puritans thought that their wealth was a sign of being chosen as the object of God's love.[85] Wealth was a sign that one was elected for salvation. In orthodox Calvinist thought, there was no other way to estimate one's 'state of grace.' A morally righteous life earned one material wealth. Hard work, 'innerworldly ascesis,' patience, economy, and modesty were considered wealth-provoking virtues and were signs of obedience to God. Weber's widely disputed analysis makes one thing undeniably clear: The enormous impact of ideology and religious thought on society and economy is incontestable. The study of economy should start with a 'deep economic' analysis, for it acknowledges the interconnectedness of economy, morals, and religion.

NOTES

1. One of the most striking historical examples of Trinity's dominant religious and cultural position is Islam's attack on this principle, calling it a tritheist betrayal of the original monotheist religion. Assuming that religion is one of the strongest systems of social legitimation, we see cultural differences between Christian and Islamic culture mirrored in this reproach. Contrary to Islam—which requires total surrender and dedication to the one God, Allah, revealed by his prophet and his successors—Christianity offers more possibilities for human autonomy and the critique of power. Here, religion symbolizes and mirrors the social system in its rigidity or flexibility.
2. For ideological reasons, I use the masculine form. I'm writing on the ideological character of the Father-Son relation in the Divine Being.
3. Its symbolic importance is comparable to models used in psychoanalysis, such as the Oedipus complex and the phallus symbol. A recent contribution to this field is Howard Eilberg-Schwartz's, *God's Phallus, and Other Problems for Men and Monotheism* (Boston: Beacon Press, 1994).
4. See, for instance, Hosea 11:1: "When Israel was a child, I loved him, and out of Egypt I called my son."
5. Dietrich Bonhoeffer proposes the expression "Christus als Gemeinde existierend" (Christ existing as community) in his dissertation *Sanctorum Communio* (1930).
6. Sigurd Bergmann, *Geist, der Natur befreit* (Mainz: Grünewald, 1995), shows how the Trinity functioned in early Christian theology even as a cosmological concept. Eastern Orthodox Christianity continued to support this tradition, stressing humans' and nature's relations, including moral principles. Bergmann doesn't hesitate to link nature's relationality and God's sociality with "the mature tradition of Stoic philosophy" (p. 122). The concept of the Trinity seems to be open to anything indeed. See also Vigen Guroian, *Incarnate Love: Essays on Orthodox Ethics* (Notre Dame, IN: University of Notre Dame Press, 1987).
7. A famous 'new name' for God is Adam Smith's 'invisible hand.' Recently, Francis Fukuyama, in his volume *Trust* (London: Hamish Hamilton, 1995), claimed 'trust' to be a divine power that keeps modern commercial society going.
8. My decades of research on Bonhoeffer's thought have led me to conclude that his theology and philosophy fit in this concept of relatedness between God and social reality, as originally laid out in the Old Testament. Consider Bonhoeffer's development concerning the latter aspect, which can be traced from his first to his last book: *Sanctorum Communio* (1930) and *Letters and Papers from Prison* (1943–44). The way that Bonhoeffer deals with the concept of metaphysics is especially important in this context. See Hans D. van Hoogstraten, "Bonhoeffer and the Problem of Metaphysics," in *Theology and the Practice of Responsibility: Essays on Dietrich Bonhoeffer*, edited by Wayne Whitson Floyd Jr. and Charles Mash (Valley Forge, PA: Trinity Press International, 1994), pp. 223–37.
9. Herman E. Daly and John B. Cobb Jr., *For the Common Good: Redirecting the Economy Toward Community, the Environment, and a Sustainable Future* (1989; Boston: Beacon Press, 1994).
10. Sallie McFague, *Models of God: Theology for an Ecological, Nuclear Age* (Philadelphia: Fortress Press, 1989); Sallie McFague, *The Body of God: An Ecological Theology* (Minneapolis: Fortress Press, 1993); Anne Primavesi, *From Apocalypse to Genesis: Ecology, Feminism and Christianity* (Minneapolis: Fortress Press, 1991).

---

11. Cf. Karl Löwith, *Meaning in History* (Chicago and London: University of Chicago Press, 1949), pp. 145–60. See also Eric Voegelin, *The New Science of Politics* (1952; Chicago: University of Chicago Press, 1987), pp. 111–2.

12. See Jürgen Moltmann, Trinität und Reich Gottes (München: Kaiser Verlag, 1980); in English, *The Trinity and the Kingdom*, trans. Margareth Kohl (San Francisco: HarperCollins, 1991). I use the German edition.

13. It seems quite to the point to suppose that Emperor Constantine was aiming at his own divine position when the Trinitarian formula was formulated at the A.D. 325 Nicaea Council. The term *homoousios* is from Constantine himself. See Moltmann, *Trinität und Reich Gottes*, pp. 149, 211.

14. Compare Hobbes's *Leviathan* (1651), especially chap. 12 ("Of Religion"), chap. 13 ("Of the Natural Condition of Mankind as Concerning Their Felicity and Misery"), and chap. 14 ("Of the First and Second Natural Laws, and of Contracts").

15. The importance of deism in modern Western self-understanding is shown in Charles Taylor's *Sources of the Self: The Making of the Modern Identity* (Cambridge: Cambridge University Press, 1992), pp. 248–85, 355–7. A direct relation between deism and evolutionary thought is indicated in Ian Barbour's *Religion in an Age of Science* (San Francisco: HarperCollins, 1990), pp. 178–85, especially p. 182.

16. See, for example, Adam Smith, *Theory of Moral Sentiments* (1759; Indianapolis, Indiana: Liberty Classics, 1982, photographic reprint of the 1976 Oxford University Press edition), II.ii.3.5: "In every part of the universe we observe means adjusted with the nicest artifice to the ends which they are intended to produce. . . . The wheels of the watch are all admirably adjusted to the end for which it was made." Smith warns humans not to imagine "that to be the wisdom of man, which in reality is the wisdom of God."

17. Interestingly, the doctrine of the so-called filioque (the Spirit emanating from the Father as well as from the Son) was a Western one. See Moltmann, *Trinität und Reich Gottes*, p. 195. This eventually caused the rupture between Latin Western and Greek-Russian Eastern religion and culture.

18. Cf. Al Gore's work *Earth in the Balance: Ecology and the Human Spirit* (Boston: Houghton Mifflin, 1989). In both the vice president's person and his work, politics and environmental ethics join together.

19. This distinction is from Jürgen Habermas. See his *Theorie des Kommunikativen Handelns*, 2 vols. (Frankfurt am Main: Suhrkamp, 1981), vol. 2, pp. 173–293, "Zweite Zwischenbetrachtung: System und Lebenswelt." This distinction is present in some form in all of his philosophical-sociological work.

20. The elite's role as exemplar in Western civilization through the centuries is broadly exposed by Norbert Elias in his work *Über den Prozess der Zivilisation: Soziogenetische und Psychogenetische Untersuchungen*, 2 vols. (1939; Bern: Francke Verlag, 1969).

21. Note the subtitle of Charles Taylor's abovementioned book: *The Making of the Modern Identity*.

22. Bill Devall and George Sessions, *Deep Ecology: Living as if Nature Mattered* (Salt Lake City: Peregrine Smith Books, 1985), pp. 66–7.

23. Daly and Cobb, *For the Common Good*, especially pt. 2, pp. 121–206.

24. Quoted by Devall and Sessions, *Deep Ecology*, p. 54.

25. Japanese corporate organic models and work ethics are extremely interesting. See Daly and Cobb, *For the Common Good*, pp. 299–300. For a broad overview, see K. G. van Wolferen, *The Enigma of the Japanese Power* (London: Macmillan, 1989).

26. The theme of inclusion and exclusion is very old indeed: It refers to who belongs to one's group or community and who doesn't. Nowadays, exclusion from information entails increasingly disastrous results. See Samuel Weber, "The Dawn of a New Age," paper presented at Multimedia Computing Conference, Utrecht, the Netherlands, 1991.

27. These contexts include important intellectual and political developments: concerning Aristotle, the Platonistic tradition and the rise of ancient empires (Persians, Alexander the Great), and concerning Adam Smith, Thomas Hobbes, John Locke, and the rise of European nation-states. An interesting effort to hermeneutically relate ancient empires with modern economic superpowers is made by Ulrich Duchrow, *Alternatives to Global Capitalism* (Utrecht: International Books, 1995).

28. See Emmanuel Levinas, *Totality and Infinity: An Essay on Exteriority* (Pittsburgh, PA: Duquesne University Press, 1969), p. 46: "Ontology as first philosophy is a philosophy of power.... Universality presents itself as impersonal; and this is another inhumanity." He warns about an ontology like Heidegger's: "Ontology becomes ontology of nature, impersonal fecundity, faceless generous mother, matrix of particular beings, inexhaustible matter for things." This leads to discomfort indeed: "Heideggerian ontology, which subordinates the relationship with the Other to the relation with Being in general, remains under obedience to the anonymous, and leads inevitably to another power, to imperialist domination, to tyranny.... Its [tyranny's] origin lies back in the pagan 'moods,' in the enrootendness in the earth, in the adoration that enslaved men can devote to their masters. *Being* before the *existent* ... is freedom (be it freedom of theory) before justice. It is a movement within the same before obligation to the other" (pp. 46–7).

29. They apply this term explicitly to economics; see Daly and Cobb, *For the Common Good*, pt. 1, pp. 25–117.

30. A special way of looking guides the reading of some representative fragments of authors who combine ontology, morals, economics, and metaphysics. What interests us most is the rise and development of modern Western economics, especially its ideological background. In the following, I borrow extensively from Arend Th. van Leeuwen, *De Nacht van het Kapitaal* (Capital's Night) (Nijmegen: SUN, 1985), published in Dutch only. See also Willem Hoogendijk's small book *The Economic Revolution: Towards a Sustainable Future by Freeing the Economy from Money-making* (London: Merlin Press; Utrecht: Jan van Arkel, 1991).

31. Aristotle's books *Ethics* and *The Politics* are both parts of one continuing argument. See van Leeuwen, *De Nacht*, p. 162: Political theory can be considered the counterpart of ethical theory, just as ethics is essentially political philosophy.

32. A crucial contradiction arises in Aristotle's reasoning concerning the slave in the spheres of the household and public life. See, for example, *The Politics* (Penguin Books), I.ii and I.iii (1253a18–1253b14): the transition from state to household. This often-neglected contradiction proves to be the obvious hermeneutical key to understanding modern efforts to defend the market-oriented economy's natural character.

33. Van Leeuwen, *De Nacht*, p. 172.

34. Aristotle, *The Politics*, 1252aI. Adhering to the original Greek text, the English translation has been altered slightly.

35. Hannah Arendt, in *The Human Condition* (Chigaco: University of Chicago Press, 1958), pp. 196–199, emphasizes that the polis is the male condition, which had to be protected by laws and walls: "Wherever you go, you will be a *polis* ... [it

is] the space where I appear to others as others appear to me, where men exist not merely like other living or inanimate things but make their appearance explicitly." Arendt, however, does not mention the special, hidden position of slaves in the public life of the polis. On the contrary, she speaks of "chances for everybody."

36. Van Leeuwen, *De Nacht*, p. 228.

37. Equality issues are pertinent to modern times, as the social contract theories of Hobbes, Locke, and Rousseau make clear.

38. Cf. Arendt, *The Human Condition*. Thus, one could speak of deep politics in Aristotle, whereas Adam Smith represents deep economy.

39. Aristotle, *The Politics*, 1295a36–1296a21.

40. See also chapter 2.

41. This is what Hegel called *das bürgerliche Denken* (the bourgeois thinking) and *die bürgerliche Gesellschaft* (the bourgeois society). John Rawls uses freedom and equality as basic ideas. See his *Political Liberalism* (New York: Columbia University Press, 1993): "Since we start within the tradition of democratic thought, we also think of citizens as free and equal persons. The basic idea is that in virtue of their two moral powers (a capacity for a sense of justice and for a conception of the good) and the powers of reason (of judgment, thought, and inference connected with these powers), persons are free. Their having these powers to the requisite minimum degree to be fully cooperating members of society makes persons equal." This all takes place "in a fair system of social cooperation" (pp. 18–9).

42. See also chapter 2.

43. Thomas Hobbes, *Leviathan, or the Matter, Forme and Power of a Commonwealth Ecclesiastical and Civil* (Oxford: Basil Blackwell, n.d. [1651 ed.]), p. xviii.

44. Ibid., p. 112.

45. Ibid., p. 164.

46. John Locke, *The Works of John Locke, Including an Essay on the Human Understanding, Four Letters on Toleration, Some Thoughts on Education, and an Essay on the Value of Money*, new ed. (London and New York: Ward, Lock & Co., n.d.), p. 560.

47. Ibid., p. 562.

48. Ibid., pp. 565–6, 572.

49. Smith, *Theory of Moral Sentiments*, III.2.32.

50. Adam Smith, *The Wealth of Nations*, (1776; Chicago: University of Chicago Press, 1976), I.iv.2. Parallels with Hobbes and Locke may be evident.

51. Cf. Kant's categorical imperative: You should behave in a way you think all others should treat you.

52. Smith, *Theory of Moral Sentiments*, I.i.I, I.i.II, I.iii.I.

53. Ibid., III.1.2.

54. Smith, *Wealth of Nations*, I.ii.2.

55. Smith, *Theory of Moral Sentiments*, I.i.I.1.

56. Van Leeuwen, *De Nacht*, p. 74.

57. Smith, *Theory of Moral Sentiments*, I.i.5.5.

58. Van Leeuwen, *De Nacht*, pp. 75–6.

59. Ibid., p. 96.

60. Smith, *Theory of Moral Sentiments*, III.ii.III: "But, in order to attain this satisfaction, we must become the impartial spectators of our own character and conduct. We must endeavour to view them with the eyes of other people, or as other people

are likely to view them. When seen in this light, if they appear to us as we wish, we are happy and contented. But it greatly confirms this happiness and contentment when we find that other people, viewing them with those very eyes with which we, in imagination only, were endeavouring to view them, see them precisely in the same light in which we ourselves had seen them. Their approbation necessarily confirms our self-approbation."

61. Smith, *Wealth of Nations*, I.iv.II: "In order to avoid the inconveniency of such situations, every prudent man in every period of society, after the first establishment of the division of labour, must naturally have endeavoured to manage his affairs in such a manner, as to have at all times by him, besides the peculiar produce of his own industry, a certain quantity of some commodity or other, such as he imagined few people would be likely to refuse in exchange for the produce of their industry."

62. Ibid., I.vi.17.

63. Smith, *Theory of Moral Sentiments*, IV.i.x: "The rich only select from the heap what is most precious and agreeable. . . . They are led by an invisible hand to make nearly the same distribution of the necessaries of life, which would have been made, had the earth been divided into equal portions among all its habitants." See Smith, *Wealth of Nations,* IV.ii.ix, for a reference to maximization.

64. Smith, *Wealth of Nations*, I.ix.xi.

65. Ibid., I.ix.xx.

66. Ibid., I.vi.xxiii.

67. *Wealth of Nations* begins with: "The greatest improvement in the productive powers of labour . . . seem to have been the effects of the division of labour" (I.i.i).

68. See also chapter 2.

69. Van Leeuwen, *De Nacht*, pp. 332–9.

70. See chapter 3.

71. See Karl Polanyi, quoted in Devall and Sessions, *Deep Ecology*, p. 136. He refuses to accept that labor, land, and money could possibly be commodities.

72. In a different way, communism shows religious features as well.

73. See the Brundtland Report: The World Commission on Environment and Development, *Our Common Future* (Oxford and New York: Oxford University Press, 1987).

74. Daly and Cobb, *For the Common Good*, pp. 407–42 ("Afterword: Money, Debt, and Wealth"). The authors include John Ruskin's aphorism: "That which seems to be wealth may in verity be only the gilded index of far-reaching ruin" (*Until This Last*, 1862).

75. 'Capital's night' coined by Van Leeuwen.

76. See Franz Hinkelammert, "The Economic Roots of Idolatry: Entrepreneurial Metaphysics," in *The Idols of Death and the God of Life: A Theology*, edited by Pablo Richard et al. (New York: Maryknoll, 1983), pp. 165–93, especially 165–8 for further examples of bodily and medical metaphors.

77. See McFague, *The Body of God*, pp. 27–63, 78–97.

78. See also chapter 5, about changing perspectives in aesthetics.

79. My thesaurus explains 'credit' as 1. trust in future payment; 2. belief, confidence; see faith.

80. Daly and Cobb, *For the Common Good*, pp. 402–42. This concept is extensively discussed in chapter 2.

81. See Bruce Rich, *Mortgaging the Earth: The World Bank, Environmental*

Impoverishment, and the Crisis of Development (Boston: Beacon Press, 1994). What happens in World Bank policy is representative, in an enlarged model, of the movement of capital in an upward spiral.

82. See John Kenneth Galbraith, *The Culture of Contentment* (Boston: Houghton Mifflin, 1992).

83. On mimesis and mimetic desire in modern literature, see Christopher Prendergast, *The Order of Mimesis: Balzac, Stendhal, Nerval, Flaubert* (Cambridge: Cambridge University Press, 1986); see also Rene Girard, *Mensogne romantique et vérité romanesque* (Paris: Editions Bernard Grasset, 1961), and Gunter Gebauer and Christoph Wulf, *Mimesis: Culture, Art, Society* (Berkeley, Los Angeles, London: University of California Press, 1995).

84. On the philosophical level, Emmanuel Levinas's philosophy of the other's face offers some exciting possibilities. In his work, the consequences of exploiting and excluding the other are in the foreground. In all Levinas's works, the meaning of the other is important. In chapter 5, I refer to Levinas, *Totality and Infinity*.

85. Max Weber, *The Protestant Ethic and the Spirit of Capitalism* (1905; New York: Charles Scribner's Sons, 1958). See also R. H. Tawney, *Religion and the Rise of Capitalism* (New York: Harcourt Brace, 1926).

Chapter 2
T w o

DEEP ECOLOGY
AND THE COMMON GOOD

THE *LOGOS* AND THE *NOMOS* OF THE *OIKOS*

Dwelling is not primarily inhabiting but taking care of and creating that space within which something comes into its own and flourishes. Dwelling is primarily saving, in the older sense of setting something free to become itself, what it essentially is. . . . Dwelling is that which cares for things so that they essentially presence and come into their own.[1]

THE FIRST CHAPTER MADE it clear that ecology and economy are affiliated concepts. Semantically, however, it is helpful to recall the original meaning of both terms. The *oikos* is the dwelling house. It should be suited for dwelling in the Heideggerian sense, and the people living in it should keep its character alive. Yet many interpretations still include "setting something free to become itself, what it essentially is."

'Ecology' concerns the *logos* of the *oikos*; 'economy' concerns the *nomos* of the *oikos* (some prefer the derivation from the word *nomeus*, meaning 'ruler,' 'shepherd'). The word *logos* is a rich concept. In ancient Greek culture, it meant 'creative power,' but it also had connotations of 'reason.' *Logos* is often used in a dialectical relation with *muthos*. *Logos* is the overcoming (in German, *Aufhebung*) of *muthos*; the *logos* interprets and exceeds the *muthos*.[2] *Logos* tries to understand the truth of phenomena in a rational way, whereas *muthos* tells unreasonable stories in which the truth is embedded. The Greek Age of Enlightenment, especially the fourth century B.C., can be characterized as the age of *logos*.

'Ecology' can be translated as the human knowledge of nature. Ecology refers to a consciousness of connectedness with nature, rather than an awe or fear of nature as a mythological realm. Humans consider themselves to be part of nature; nature is their home (*oikos*). Since humans now know about biospheres and food chains, and hence about the connectedness of all life, they should maintain their home in a responsible and respectful way. In this instance, the *logos* is not human reason in the dualistic, Cartesian sense of the word, which places an absolute priority on reality as dependent on human reason. In fact, most deep ecologists interpret the (eco-)*logos* as the power of an ontological order that inhibits reality. When humans willingly consider themselves to be a part of this ontological order, they enable themselves to understand it. In this case, human reason does depend on reality.

So even if it is not explicitly formulated, the deep ecologist's use of the *logos* is closely related to the original interpretation of the word. In the term 'ecology,' *logos* not only refers to modern, empirical knowledge of something but also includes connotations of metaphysical knowledge, insight, and wisdom. Thus, 'ecology' asks not only the empirical question how does it work but also the metaphysical question what is the *oikos'* origin and being? And like all natural philosophy and natural theology, the *logos* of the *oikos* carries a moral imperative. This moral imperative simply means that humans, acknowledging and accepting their position in the natural system, should live in accordance with it. Deep ecology's radical wing in the United States, Earth First, exemplifies this imperative: Humans are subordinate to the earth's well-being.[3]

As discussed in chapter 1, there are evident differences between deep ecology and deep economy. In typical usage of the word, the *nomos* (economy) of the house is the law. In the New Testament, particularly in Paul's letters, the *nomos* is often interpreted in the negative sense of capturing people and coercing them to behave in accordace with fixed patterns. Gradually, the 'law' lost its character as God's Torah, and it acquired the character of something from which people should be liberated. In early ancient Greek thought, the *nomos* was rooted in religion, having a general meaning of cosmic validity. From Herodotus on, however, the *nomos* was increasingly interpreted as phenomenon, that is, as a locally and temporally different morality. This included the legal-political system. Unlike *logos*, *nomos* lost its strong ontological character; it lost the meaning of morality as such.[4]

It is important to realize that the concept of *nomos* as a human, political, and moral set of laws and rules was increasingly emancipated from religion and nature in a metaphysical-ontological sense. This development was closely connected to the desire to be autonomous. It was not until the eighteenth-century Enlightenment that the individual's autonomy emerged as an increasingly achievable state. The resulting broad emancipation during this period fundamentally changed relationships and components in the social system. 'Freedom' was the key term.

This well-known development determined modern economy's deep character. As described in chapter 1, external consequences were not its only effect; on the contrary, it changed the way people thought, and it impacted their self-determination.[5] From a historical perspective, emancipation is still in an early phase of development. And although many recognize its actual and potential dangers, few involve themselves on a critical level morally, ideo-logically, scientifically, or even practically.

The establishment of the 'free'-market economy is one of emancipation's great deeds. We could call it the Western white male's victory over natural limitations. Many modern people hold a deep conviction that this is the achievement of a victorious culture. (This is not necessarily an ideological statement, although I consider it one.) The rupture that Western culture provoked in human history is dialectical in character.[6] As the old religion was done away with, a new religion rose in its place. However, before we can name this new religion, we must analyze and understand emancipation's impact. Moreover, before changes in the economic and political fields are proposed, we must first understand the historical processes, including theoretical legitimations, that established the free-market economy and democratic politics.

It can be assumed that people want to live comfortably in their 'house.' The economy informs them of the laws and rules that they must obey, albeit as free, emancipated citizens of democracy, in order to sustain their 'house.' In 'a world come of age,' humans live with the idea that they are responsi-ble for themselves and their life conditions.[7] There is also a deep conviction that life on earth for humans is possible only if certain conditions are met. This unique combination gave *chrematistics* an opportune developmental chance. However, the basic law of *chrematistics*, free enterprise in various fields, does not parallel nature's *logos*. Therefore, modern people are not 'naturally' adapted to nature, consequently, they feel alienated from it. As we have already seen, this *nomos* is based largely on self-interest. Humanity's self-centeredness makes it impossible to accept a modest position among other species. It also inhibits humanity from addressing the moral consequences of its practices.

HOW DEEP IS DEEP ECOLOGY?

George Bradford wrote a small critical book entitled *How Deep Is Deep Ecology?*.[8] His main argument is similar to Daly and Cobb's study,[9] in that he suspects the deep ecologists of being contemptuous toward the human in their emphasis on earth and species. His point can be reduced to biocentrism versus anthropocentrism.[10] It is an effort to clarify the most basic pre-suppositions. If professionals cannot agree on this epistemological point of departure, any argument concerning 'deep' or 'shallow' economy is nonsen-sical. Bradford states:

> Deep ecologists err when they see the pathological operationalism of in-dustrial civilization as a species-generated problem rather than as one generated by social phenomena that must be studied in their own right. Concealing socially generated conflicts behind an ideology of 'natural law,'

they contradictorily insist on and deny a unique position for human beings while neglecting the centrality of the social in environmental devastation. Consequently, they have no really 'deep' critique on state, empire, technology, or capital, reducing the complex web of human relations to a simplistic, abstract, scientistic caricature.[11]

In their 1989 'great debate,' Murray Bookchin and Dave Foreman discussed this precarious epistemological problem.[12] These representatives of social ecology (Bookchin) and of deep ecology/Earth First (Foreman) offer hope that discussion is possible. Certainly, they will not accept a deadlock, which is how one could describe Bradford's critique. Obviously, then, it is necessary to discuss biocentrism versus anthropocentrism on a fundamental level. The analysis of deep economy is a core part of this discussion. The central question is why the anthropocentric approach toward reality has a devastating effect. The content of Western culture's anthropocentrism is at stake.

After criticism that their views and prescriptions were rather vague, several deep ecologists clarified their themes and appealed to their thinking audience: "Readers are encouraged to elaborate their own versions of deep ecology, clarify key concepts and think through the consequences of acting from these principles."[13] We should refamiliarize ourselves with the basic points of well-known studies. In their book on deep ecology, Devall and Sessions introduce the Norwegian philosopher Arne Naess, a prominent advocate of an organic approach of nature. Summarizing fifteen years of thinking, Naess and Sessions, in 1984, formulated eight basic principles of deep ecology:

1. The well-being and flourishing of human and nonhuman Life on Earth have value in themselves (synonyms: intrinsic value, inherent value). These values are independent of the usefulness of the nonhuman world for human purposes.
2. Richness and diversity of life forms contribute to the realization of these values and are also values in themselves.
3. Humans have no right to reduce this richness and diversity except to satisfy *vital* needs.
4. The flourishing of human life and cultures is compatible with a substantial decrease of the human population. The flourishing of nonhuman life requires such a decrease.
5. Present human interference with the nonhuman world is excessive, and the situation is rapidly worsening.
6. Policies must therefore be changed. These policies affect basic economic, technological, and ideological structures. The resulting state of affairs will be deeply different from the present.
7. The ideological chance is mainly that of appreciating life quality (dwelling in situations of inherent value) rather than adhering to an increas-

ingly higher standard of living. There will be a profound awareness of
the difference between big and great.
8. Those who subscribe to the foregoing points have an obligation directly
   or indirectly to try to implement the necessary changes.[14]

Sessions and Devall have also presented a scheme that identifies the contrast
between the dominant worldview and deep ecology:

Dominant Worldview	*Deep Ecology*
Dominance over Nature	Harmony with Nature
Natural environment as resource for humans	All nature has intrinsic worth/biospecies equality
Material/economic growth for the growing human population	Elegantly simple material needs (material goals serving the larger goal of self-realization)
Belief in ample resource reserves	Earth "supplies" limited
High technological progress and solutions	Appropriate technology; nondominating science
Consumerism	Doing with enough/recycling
National/centralized community	Minority tradition/bioregion[15]

In this vision, biocentrism has absolute priority. Devall and Sessions want
to overcome the dominant worldview. There is a difference in terminology
between my work and that of Devall and Sessions. What I call 'deep economy,'
they call 'dominant worldview.' When I use the term 'deep economy,' I
mean reflection on and analysis of the hard materialistic reality that is supported by a general belief in human progress, emancipation, freedom, and
autonomy. The deep ecologists call this a 'worldview,' and in opposition to
it, they try to formulate a different, more acceptable one. Their preference
is for a worldview akin to a premodern type of society. They assert that
human existence is not superior to the existence of other species; according
to biocentrism, all life-forms are equal.

   In *For the Common Good: Redirecting the Economy Toward Community,
the Environment, and a Sustainable Future*, Herman Daly and John Cobb
Jr. make a serious start toward realizing their title. They accurately describe the ecological and economic spheres. In their study, the issue of biocentrism is important. In response to Naess and Sessions's eight principles,
they write:

For Naess and Sessions our position is in fact excluded from deep ecology despite our acceptance of the eight basic propositions. For them, basic points 1 and 2 are interpreted in terms of "biocentric equality," the intuition "that all things in the biosphere have an equal right to live and blossom and to reach their own individual forms of unfolding and self-realization within the larger self-realization . . . that all organisms and entities in the ecosphere, as parts of the interrelated whole, are equal in intrinsic worth." . . . We do not share this view. We believe there is more intrinsic value in a human being than in a mosquito or a virus. We also believe that there is more intrinsic value in a chimpanzee or a porpoise than in an earthworm or a bacterium. *This judgment of intrinsic value is quite different from the judgment of the importance of a species to the interrelated whole. The interrelated whole would probably survive the extinction of chimpanzees with little damage, but it would be seriously disturbed by the extinction of some species of bacteria.* We believe that distinctions of this sort are important as guides to practical life and economic policy and that the insistence that a deep ecologist refuse to make them is an invitation to deep irrelevance.[16]

Daly and Cobb make an important point concerning deep ecologists' persistence regarding value and equality ideology, which actually invites 'deep irrelevance.' Deep ecologists admit that species use one another for food, shelter, and so forth, but they simultaneously forward their "basic intuition" of radical equality, achieved by "turning attention away from these practical issues."[17]

In my opinion, the answer is not difficult. Fascinated by the earth's richness, these biocentric deep ecologists support all species by giving them a human qualification of value. For decades, natural theology and philosophy have been speaking of 'intrinsic' value. This term seems to set natural species free from the value that humans award them. The economic and aesthetic interpretation of value is considered inferior to nature's own creational value. Creational value is called 'intrinsic.' People who propose this qualification, however, forget that they themselves are qualifying subjects. Of course, the philosophical problem that now arises is whether the human subject can transcend his or her own context: Can the human 'leap over his own shadow'? To date, the evidence is not in favor of human metaphysical arobatics; in fact, it seems that the attempt has a negative turn: 'Intrinsic' value becomes estimated in human terms of inferiority or superiority. History has proved that Superman (*Der Uebermensch* of Friedrich Nietzsche) presupposes the other's inferiority.[18]

Daly and Cobb make important contributions to the deep ecology discussion, but they falter when they neglect to consider the value argument's validity. Instead, they focus on an alternative line of thought, biospherical connectedness. Their approach in this area consequently calls their solution of emphasizing the interrelatedness of species into question. As Bradford states:

The deep-ecology perspective insists that everything is interrelated and sees this recognition as "subversive to an exploitative attitude and culture" (Sessions, in Tobias anthology). But ecological reductionism fails to see the interrelatedness of the global corporate-capitalist system and empire on the one hand, and environmental catastrophe on the other. . . . In fact, the absence of a critique of capital is a real impediment to the generalization of authentic resistence to the exploitive-extractive empire which is presently devouring the earth, because it mystifies the power relations of this society and squanders the possibility for linking the human victims of the machine in different sectors.[19]

I support Bradford's argument regarding deep economy, because it shows an awareness of modern ideological practices. His intuition is striking. In discussing William Catton's *Overshoot*, which he judges as highly ideological, he notes: "Attempting to turn 'ecological principles into sociological principles,' he turns sociological distortions into natural law."[20] Thus, Catton follows Adam Smith's lead—he combines moral philosophy with modern economy.[21]

Daly and Cobb reject the concept of intrinsic value, especially forced equality, in favor of 'interrelatedness' as a moral and critical standard with the potential of redirecting the economy toward community, the environment, and a sustainable future. Thus they partly support deep ecology. Their presupposition is that when people are convinced of the basic structure of human existence, which is community, they will be able to overcome economy's destructive nature. According to Daly and Cobb, the realized community will practice economics in its original, serving sense.

COMMUNITY AS A SAVING CONCEPT

'Misplaced concreteness' is one of Daly and Cobb's central themes. They derived the term from Alfred North Whitehead, who spoke of "the fallacy of misplaced concreteness" regarding the modern university's division of knowledge.[22] Daly and Cobb contrast abstractness and concreteness as critical hermeneutical tools for the judgment of theoretical and practical positions. One could evaluate their treatment of these concepts as the project's core argument. For this reason alone, it is worthwhile to look closely at their argument.

Thinkers often forget the degree of their thought's abstractness and proceed to declare their work concrete. Thus 'abstractness' is identified as 'misplaced concreteness.'[23] The field of economics is overrun with instances of this. Therefore, we must ask whether the theorist's stress on misplaced concreteness is essentially the same discussion as the one about deep economic structures. The answer is no. A representative quotation brings us quickly to the heart of the matter:

What is the set of abstractions that political economy has riveted on economic thought and at which it has come to a self-satisfied halt? One of
the most important is the abstraction of a circular flow of national product and income regulated by a perfectly competitive market. This is conceived as mechanical analog, with motive force provided by individualistic
maximization of utility and profit, in abstraction from social community
and biophysical interdependence. What is emphasized is the optimal allocation of resources that can be shown to result from the mechanical inter-
play of individual self-interests. What is neglected is the effect of one
person's welfare on that of others through bonds of sympathy and human
community, and the physical effects of one person's production and consumption activities on others through bonds of biophysical community.[24]

Here, Daly and Cobb mention "individualistic maximization of utility and
profit, in abstraction from social community and biophysical interdependence." Apparently, the most concrete social phenomena are the social community and the biophysical interdependence. We can think of this as the
bonds of sympathy and human community and the bonds of biophysical
community.

According to the text, current economic exchange relations are abstracted
from the human's condition as a social being. The motivating force is located in "individualistic maximization of utility and profit." Daly and Cobb
claim that this is the way it happens, and they condemn it as an abstraction
that causes a fatal dichotomy. Their normative ontological concept of reality
is the creational bond between people and between humans and nature. They
advocate that economy and politics need to adhere to the original relational
structures that exist between human and nonhuman life. The problem arises
when Daly and Cobb subsequently identify an individualistic, egoistic motive force. This motive force must incline the individual to place a distance
between the self and the community. Such a distance, however, is a 'sin'
against the original position.[25]

What is the basis for Daly and Cobb's surety regarding the good community versus the bad individual? The only feasible answer is a hidden
deontological ethical position in which the 'is' precedes and dictates the
'ought.' Because people are living in communities and because virtues such
as sympathy are dominant communal features, they should practice an economics (like the Aristotelian *oikonomia*) that serves and strengthens the original
community. People cannot and should not leave the original bonds. As social beings, their lives depend on strengthening and sustaining these bonds.
The problem with this kind of reasoning is that economics is approached
with moral arguments. Using the original position as a point of departure,
all diverging theories and practices are warned and challenged to conform
in the right way—to adapt to the original position.

As I argued in the first chapter, 'deep' economy thinkers such as Adam Smith defend the opposite position. Smith deals with the free-market system as a natural datum that includes exchange relations as a natural source of the common wealth. Virtues, such as sympathy, function as the market's stimulating moral force, and they link egoism and altruism. Virtues provide the exchange process with a foundation of trust, a vital condition for the market's original task: providing wealth and achieving global well-being. Satisfying self-interest eventually builds communal well-being. Similar to Daly and Cobb, however, Smith concludes with a warning: People who will not cooperate injure themselves and the community.

The problem with both Smith's and Daly and Cobb's approach is their abstraction from the factual, historical, and social reality. Defining communal bonds as the concrete absolute may be anthropologically and morally correct, but it is not historically correct. Both approaches are reading reality in an unhistorical and hence unrealistic manner. Both approaches are therefore abstract rather than concrete. This is the reason that Daly and Cobb's abstract versus concrete discussion never arrives at a point of solution. They are literally caught in the discussion. Communal bonds may very well be the concrete absolute, but if this is so, it must be shown on the level of fact, history, and reality, not simply on a theoretical level. Presently, communal bonds can make no such claims for concreteness.

On this point I want to be perfectly clear. In the Western type of society, which is rapidly conquering large parts of the world, all types of primeval/original/ideal communities are gone, with the exception of certain museum-like places. We cannot simply reawaken these lost communities; the result inevitably will be either a harmless or a harmful idealistic concept.[26] Emancipation, communication, technology, secularization, and the market economy have all contributed to the human community's disappearance as a historical reality.

We can refer to this as the process of 'modernization.' Postmodern society's recent loss of communal and individual identity is a result of modernization. In postmodern society, the most protected and cherished social institutions and traditions are called into question. Where does the family-community stand? What do we think about domestic abuse? What is the position and importance of the church-community? What about the state-community? What do we think of its class character and prejudices? What about bureaucracy and politicians' private interests? I am aware that it is possible to have a positive approach. But in my opinion, all grounds for awarding the human community a moral-metaphysical status have been lost.

Community cannot function as a 'saving concept,' at least not in the sense of an anthropologically or biologically founded postulate. We should not give community the qualification of a given ontological structure. The reason for this is that in modern times, individuals form communities before communities form individuals. Human community is still a vital concept. But the 'saving chance' of community is a future reality—a community that people must form. Even though humans are social animals, community is not a given ontological or biological fact.

One of modern economy's concrete consequences is the suffering other. I use the term 'concrete' because the suffering other's existence is a hard, historical fact. By 'other,' I specifically mean the victims of economic growth, humans as well as nature. Searching for the mechanisms of exploitation and exclusion that cause this suffering seems to be a challenge. This concreteness involves a personal challenge, rather than a challenge to a presumed, postulated community. People who analyze and fight concrete disease-provoking relations could form bonds that create a community, but the bonds must precede the community. In order to start 'community development' in an entirely new way, we need courage, faith, hope, and love—the building blocks of bonds.

MODERNIZING ARISTOTLE'S POLIS

Continuing the rough outline in chapter 1, I return once again to Aristotle's idea of the polis as the original community. As I will show, Daly and Cobb try to found their community theory in this deep political thinking, thus missing the chance to pay enough attention to the fundamental rupture that runs between ancient and modern Western social thought—between deep politics and deep economy.

Historical excursion helps localize the semantic field we are working in, and it also brings us to the semantic ground of our intellectual forebears. Moreover, it gives our arguments a background that transcends the contemporary voice. In our effort to gather the building blocks of a safe, new house (*oikos*), we cannot ignore our own tradition. Yet there have been many people who have tried to build a safe *oikos* for themselves in history. Since we have some knowledge of the fate of their society, we might be able to learn from their experiences—successes and failures. In appealing to the past, however, we must always perform the hermeneutical task of assessing our historical representative's cultural and philosophical background. When we compare and interpret texts, we are working on a semantic field. Semantics, however, cannot be removed from its historical and social context. We should keep this in mind so that our investigation of language's meaning and our subjective attribution of textual meaning will be responsibly improved from a semantic standpoint. I begin the historical review by further considering Aristotle's polis, which casts an illuminating light on Daly and Cobb's concept of community.[27]

Daly and Cobb describe Aristotle's concept of the polis as the real community in which economy's misplaced concreteness is banned. They deal with the problematic relationship between *chrematistics* and *oikonomia*.[28] As I did in the first chapter, they highlight the community character of Aristotle's polis and his rejection of *chrematistics*. They correctly assess that Aristotle's economic position is the opposite of modern economics. Aristotle's *oikonomia* excludes *chrematistics* as highly unnatural. As previously discussed, this means a totally different place and function of money. The polis's organization forbids accumulation, because it would harm the common good.

Analyzing the differences between *oikonomia* and *chrematistics*, Daly and Cobb write:

> Oikonomia differs from chrematistics in three ways. First, it takes the long-run rather than the short-run view. Second, it considers costs and benefits to the whole community, not just to the parties to the transaction. Third, it focuses on concrete use value and the limited accumulation thereof, rather than on abstract exchange value and its impetus toward unlimited accumulation. Use value is concrete: it is a physical dimension and a need that can be objectively satisfied. Together, these features limit both the desirability and the possibility of accumulating use values beyond limit. By contrast, exchange value is totally abstract: it has no physical dimension or any naturally satiable need to limit its accumulation. Unlimited accumulation is the goal of the chrematist and is evidence for Aristotle of the unnaturalness of the activity. True wealth is limited by the satisfaction of the concrete need for which it was designed. For economia, there is such a thing as enough. For chrematistics, more is always better.[29]

This quotation exemplifies Daly and Cobb's moral approach toward economics. They describe the 'chrematist' as an inferior moral subject or agent. The 'chrematist' should adhere to moral standards, such as "the wealth is limited by the satisfaction of the concrete need" and "there is such a thing as enough." Their moral argument is located in an economic structure: *Oikonomia* "focuses on concrete use value and the limited accumulation thereof, rather than on abstract exchange value and its impetus toward unlimited accumulation. Use value is concrete: it is a physical dimension and a need that can be objectively satisfied." Thus, Daly and Cobb apply the modern distinction of use and exchange value to the ancient concept of *oikonomia*. In modern thought, this involves a technical distinction concerning free-market economy, competition, growth, and progress. Daly and Cobb, however, suggest that Aristotle succeeded in 'taming' exchange value on the basis of 'steady-state economics' or an 'economy of enough.' To clarify this rather confusing mix of ancient and modern terms, we need to recall Aristotle's 'closed' worldview.

The *oikonomia* is part of Aristotle's politics; it functions in ruling the polis. Politics and ethics are both part of metaphysics. Thus, in Aristotle's thought, politics is 'natural' and hence 'moral.' Acting according to nature is morally right, and vice versa. This is part of Aristotle's metaphysical concept of reality. In this concept, free male citizens are morally responsible for the polis's right functioning. The other beings, such as the slaves, are subordinate to the lord's moral agency. The slaves' responsibility is limited to obedient cooperation with the lord. The slave has no dealing with money whatsoever. In Aristotle's vision, *chrematistics* would disturb the hierarchical, natural, and metaphysical order of the community. *Oikonomia* is the community's *nomos*, the natural law that establishes the good household. In Aristotle's social construction of reality, all members of the polis feel at home. Feeling at home means accepting and feeling comfortable in one's position. As we can see, Aristotle's concept of the polis community is fundamentally different from our own socioeconomic community. The modern, 'emancipated' meaning of economy's *nomos* and Aristotle's *oikonomia* are not comparable concepts.

All this avenges itself on the hermeneutical field. Daly and Cobb's neglect of Aristotle's 'deep' analysis of community results in a shallow ancient-modern comparison. Haphazardly, they apply their standard of concreteness to the ancient concept of community and then proclaim it, via unhistorical methods, the model for modern community. In contrast, Aristotle's thoughts are based on hard, concrete reality: There are human beings and slaves. The slaves help the male humans to be human, but they are not human themselves. Aristotle's moral and political theories are built on this presupposition. The Athenian superman presumes that his slave counterpart is a necessary but inferior being, at least compared with him, the master. By not digging far enough into the 'deep' economy of ancient community, Daly and Cobb invalidate their proclamation that Aristotle's polis is a working model for our type of modern community. This does not mean, however, that we should throw out their main point: the opposite valuation of *oikonomia* and *chrematistics*. On the contrary, awarding *chrematistics* a divine status is a tragedy in modern history, and tracing the period in which ethics and economics underwent this catastrophic change is a worthwhile endeavor and a valuable contribution.

Homo Naturalis and Homo Economicus

The integration of the ancient and Christian worldviews began with the spreading of Christian culture by the Roman empire. During the Middle Ages, Aristotelian thought still reigned, but it was adjusted to the Christian religious system. The late Middle Ages witnessed major changes in almost all segments of society. The human spirit was not immune to these changes.

The path of development initiated and forged by these changes continued through the Renaissance and is still discernible in the present. This path marks the rise of bourgeois society (*die bürgerliche Gesellschaft*). Although different European countries experienced this rise and development during various historical eras, the entire process is known today as "modernity."

The rise of bourgeois society overran the traditional experience of community. The pyramid concept of social classes that flourished during the Middle Ages was gradually leveled to a society of equal individuals. The abstract concept of equality was made concrete by the individual's establishment of his or her own position. There was no longer any room for the so-called natural community. 'Birth and blood' gave way to financial status as the determining factor for societal position. Successful bourgeois individuals rose in social position according to their ability to accumulate wealth. The bourgeois flourished as society (*Gesellschaft*) became a place in which individuals could freely enter into a 'social contract.'[30]

Bourgeois (*bürgerliche*) status was made possible by neutral means—money—not by the traditional community. The unlimited creation of money and the simultaneous accumulation of capital were considered necessary for societal development. All people must have the opportunity to make the abstract possibility concrete in their own lives. This may very well be the most important notion underlying the grand-scale societal change. Groups, positions, and communities are no longer members of a static social reality but are dynamic agents of a society in progress. Of course, people remain in certain communities, such as the family, but these also move from lower to higher positions (and vice versa). Moreover, in their new environment, people form new communities of friends, colleagues, business partners, clients, networks, and so forth. The dynamic quality of modern society is made possible by the modern concept of society. Thus, we can see why Daly and Cobb's 'community of communities' is highly questionable.

Wealth has always functioned as an indicator of class and status in Western civilization. The difference between modern and premodern times is the availability of the sources necessary for acquiring wealth. There is a direct parallel between the demand for equality and the free-market economy (including *chrematistics*).[31] Everyone, which initially meant only white males, should be able to provide his family with the means to climb the social ladder. The means for this upward journey is money. Money is the acknowledged and accepted means of exchange. Money as the means of exchange applies to the acquisition of property as well. But how does one lay one's hands on this money? The possibility lies in the exchange value of commodities. As long as utilities have a specified use value, the accumulation of wealth is impossible, especially at the societal level.

In this historical development, there is a causal connection among the erosion of original human communities, the shifting social pattern, and the 'economic New Deal,' *avant la lettre*. *Homo economicus*, the new Western human, loses his ancestral ground, and his identity as *homo naturalis* disappears along with it. The meaning of nature completely changes. The ancient concept of nature's metaphysical and moral security and consistency vanishes. Natural laws are no longer considered imperatives for human behavior. Science and technology investigate nature and its laws in the interest of manipulating them to service human comfort. This indicates a radical new conception of the meaning of 'natural law.'

Did *homo economicus* lose his capacity as *animal sociale* or *zoon politicon*? Ecologists blame modern ideology for losing humanity's final truth: being a part of the biospheric universe. Ecologists award this final truth a vital power that enables the lost people to find their way back home. Ecology is superior to economy in this sense. The concept of community seems to be the most obvious mediating path for *homo economicus*'s return to *homo naturalis*, from alienation to the reclaiming of the self.[32] Connectedness and relatedness constitute the magic formula.

Daly and Cobb join these community-advocating ecologists. However, they cannot entirely abandon *homo economicus*, because they want an economy that serves the community. They claim the writings of Ferdinand Tönnies, a nineteenth-century scholar, to explain their trust in the concept's vital power.[33] Daly and Cobb agree with Tönnies's distinction of *Gemeinschaft* (community) and *Gesellschaft* (society), but they reject the permanence of community and society as two types of human groupings, one intimate and the other impersonal. Instead, Daly and Cobb view community as one form of society, and they claim that its establishment is possible by awarding community these four characteristics:

1. Society's communal character does not entail intimacy. It does entail that membership in the society contributes to self-identification.
2. There is extensive participation by society's members in the decisions by which its life is governed.
3. The society as a whole takes responsibility for its members.
4. This responsibility includes respect for the diverse individuality of these members.[34]

Daly and Cobb call this type of society-community 'normative,' and they advocate political and economic decentralization. An example of the execution of these moral tasks is the Catholic teaching of 'subsidiarity,' which entails decentralizing offices and delegating them to the level of social mesostructures. Daly and Cobb advocate this in the hope that both political and moral aims will be served. They believe that society's moral capital has been depleted in the modern market due to the never-ending struggle to accumulate financial capital (p. 51).

Daly and Cobb attribute primacy and superiority to the political society-community. Similar to Aristotle's vision, economic action has to conform to this model. Designating their model as concrete, they tend to consider differing models as abstract or as 'misplaced concreteness.' The problem with this is that today's global market happens to be the most concrete system that has ever existed. This is true as long as we do not separate the market from the whole progressing system, in which the exchange of commodities as the heart and soul of the free market occurs at a relatively late stage. Because today's global market exists, any small, well-organized community is a dependent part of the whole. Being a part of the system is an unavoidable and inescapable reality.

Daly and Cobb recall ancient Hebrew culture to establish original human connectedness with the land. In this way, they identify a connectedness with nature, creation, and the environment (p. 97). This context brings property rights under prophetic critique. They incorporate the biblical prophetic God into God as the inclusive whole (pp. 389–93). Two aspects of the prophetic tradition are important: idolatry and personal responsibility. Idolatry is formally defined as "treating as ultimate or whole that which is not ultimate or whole" (p. 389). This is similar to the fallacy of misplaced concreteness, in which case something is mistaken for concrete. The Hebrew prophets showed a passion for the righteousness of Israel as a community (p. 390).[35] As long as there is no idolatry involved, Daly and Cobb sympathize with deep ecologists' holistic concepts.

According to Daly and Cobb, the process of thinking is the human result of the combination of natural and revelational knowledge: "all events, including acts of human willing, are largely the outcome of antecedent events," and "something happens afresh in each event" (p. 399). Humans are able to choose, but there are universal truths—for instance, the truth of righteousness. Daly and Cobb call it the "theocentric undergirding of the biospheric perspective."

They do not want to give up this theocentrism for several reasons. First, it checks idolatry. Indeed, it counters "the conviction so deeply rooted in the discipline that growth is both the supreme end and the supreme means for achieving the end" (p. 402). Second, God functions as a moral standard, because "the true God is the omniscient and impartial unifying source of all" (p. 403). Third, the belief in God is a necessary correction to the false belief that the earth's chemical balance constitutes the ultimate tribunal of meaning.[36] Fourth, theism "provides a basis for understanding our relation to the future." It also prohibits our inclination to concentrate merely on the present: "God is everlasting, and future lives are as important to God as present lives. To serve God cannot call for sacrifice of future lives for the sake of satisfying the extravagant appetites of the present" (p. 404).

Community is the only social gestalt that combines the practicing of virtues with being a part of a much larger totality of connectedness. Assuming that the prophetic voice of Israel was intended to keep the community pure, should the moral behavior of today's community still esteem this voice in the interest of its own purity? Is process thinking applicable to the inhabitants of modern society? These questions fuel the upcoming discussing regarding the confrontation between *homo naturalis*'s vital strength and *homo economicus*'s social power.

BIOSPHERE, ECONOMY, AND THE SUBJECT

Daly and Cobb identify a division between Adam Smith's moral-philosophical writings and his economic writings, particularly between *Theory of Moral Sentiments* and *Wealth of Nations* (pp. 159–65). They use the division between morals and economics, represented by Adam Smith himself, as a prototypical example for the development of *homo economicus*. Of course, this contradicts my own interpretation of Smith in the first chapter of this book. But before we go further with their analysis, we should review Smith's main ideas.

Smith asserts that capital accumulation happens through actions of self-interest, which is right, but these actions destroy community's morality, and this is wrong. Smith has a simple solution for this dilemma. Similar to Aristotle's use of virtue, Smith models the community system's optimal functioning on the virtues of fairness, justice, sympathy, and so forth. Smith asserts that the same is true for a well-functioning capitalist society. Production, exchange, and consumption are impossible without cooperation and a certain amount of fairness, sympathy, and trust. In the first chapter, I conclude that these virtues coexist with exploitation as the real (concrete) driving force of the system. I argue that assuming that virtues are community's concrete guidelines and declaring exploitation as a case of misplaced concreteness are simply too easy. I assert that exploitation is a plainly concrete experience, whereas community virtues are the system-sustaining ideology that frees those who are active in economic life from the responsibility of concrete reality.

Daly and Cobb attribute different meanings to concreteness and abstraction. Their standard is community or, in political terms, the 'community of communities.' To define their views on the real, concrete *homo economicus*, they consider the human to be a radical social being (pp. 159–89). Thus, they "call for rethinking economics on the basis of a new concept of *Homo economicus* as person-in-community" (p. 164). They admit that being a person-in-community "does not preclude an element of individualism," but they also do not see an actual conflict:

> But what is equally important for the new model—and absent in the traditional one—is the recognition that the well-being of a community as a whole is constitutive of each person's welfare. This is because each human being is constituted by relationships to others, and this pattern of relationships is at least as important as the possession of commodities. . . . Hence this model of person-in-community calls not only for provision of goods and services to individuals, but also for an economic order that supports the pattern of personal relationships that make up the community. (pp. 164–5)

As if slowly lifting a veil, they reveal step-by-step the fundamental, metaphysical concept of community. After describing all kinds of sociopsychological features of persons-in-community, the main argument turns out to be derived from the biosphere. Here the congeniality toward deep ecology becomes clear. As Daly and Cobb state:

> How should the biosphere be conceived? The internal relatedness of all its parts forbids thinking of it as composed of self-contained individuals as if the whole were the addition of separable units. Each of its members is social. The biosphere is a society, or rather a society of societies. (p. 202)

They admit that we cannot speak of a biospheric community, because the criteria for a community cannot be met. The presupposed levels of subjectivity are rarely found outside human societies. But suddenly there seems to be a possibility of approaching the biospheric system as a community:

> Human beings can derive part of their identity from membership in the biosphere. They can participate in decisions it makes, and they can care for the whole, as well as for its individual members in their diversity. In this qualified sense, *for its human members*, the whole biosphere can and should be a community of communities. (p. 202)

On this point, their reasoning is inconsistent. First, there is insufficient subjectivity among nonhuman species, but somehow the biosphere is making decisions in which humans should participate. This establishes a moral imperative, or at least a moral standard, that is provided by the biospheric decision maker. The strength of this imperative will eventually bring economists to their knees:

> When economists deal with the living things, and especially with large systems of living things, they cannot think of these *only* as resources for fueling the human economy. . . . We see signs of a rising biospheric consciousness in many segments of contemporary society. We hope it will soon begin to shape economic theory and practice. (p. 203)

Thus, for the time being, there are two opposing and exclusive views: the view of human subjectivity, which defines humans as moral agents (the deep economic and ecological perspective), and the opposing view of human autonomy derived from the (transcendental) self.[37] Deep ecology and deep economy both acknowledge a dependency on a larger whole or system. Neither settles for language or religious conviction. Deep ecology locates this dependency in human membership in the biospheric universe; deep economy locates it in membership in a money-producing society. Both agree that humans are engaged in life-saving activities.

The reason for this mutual exclusion might be that neither takes the other's view seriously. Deep ecologists can hardly stomach the application of their terminology to economy. For them, 'deep' economy is a contradiction in terms. This is a serious matter. As previously argued, economy's 'deep' character is an integral part of the broad enlightenment movement. Thus it cannot be separated from 'the making of the modern identity.' Economy's language, reasoning, and practice are undeniably intertwined with the Enlightenment's worldview. We should question the Enlightenment's concept of freedom in terms of 'economic housing.' We should also scrutinize the free market's divine character. These 'deep' features should not be neglected. But most strong nature-oriented thinkers and activists routinely neglect them. These thinkers and activists are quick to point the finger at economic and political effects, but they are astonishingly reluctant to excavate the roots of cause.

Economists and politicians demonstrate the same caliber of inaction regarding deep ecology thought. Politicians are hesitant to bring biosphere issues into the public eye, because economic relations are the mainstay of their constituents' practical concerns. Biosphere issues threaten economic stability and hence political favor. In economists' theoretical work, one hardly ever finds (deep) ecological reflection or even deep reflection on the discipline of economy itself. Herman Daly is the one exception, but his writings are not taken seriously by his colleagues; they may even feel that he is a traitor.[38]

What position allows serious reflection on deep ecology and deep economy? To find this position, we must solve the subject problem. My opinion is that we are first and foremost individuals who, after striving to attain freedom and autonomy, must fight to defend them. In the process, we are usually members of communities, but we are never members of a 'community of communities.' We are dichotomies, and as such, we play different roles. We are members of society as well as members of smaller communities within society; this is modernity's character. But this is not the ultimate truth about modern humanity. We are historical as well as social beings. Or, more specifically, because we are social, intelligent creatures, we are historical beings.

This means that we are members of a tradition. My concept of tradition goes far beyond the tracing of geographic or ethnic origins, as an American charts his or her European ancestory, for example. Instead, the tradition is located in the great decisions and commandments of the Moses tradition (Jesus being the second Moses). These decisive words and actions affected the relations between humans and between humans and nature, all within the semantic field of *dabar YHWH*, the Word of the Lord. The joint Jewish, Christian, and Islamic culture's inheritance of the Jewish tradition may be vital for our individual awareness of being a subject. Our conception of nature and our fellow humans as the others is derived from our consciousness of the tradition's major outcomes.

The hermeneutical problems are numerous. This prevents me from talking of a 'community of tradition' or its equivalent. In this respect, there is no preexisting community. Given the knowldedge of their inheritance of membership in a tradition, modern individuals face the task of interpreting the tradition's treasures for and with their contemporaries. Through deep discussions with fellow members of their tradition, they might discover that they do form a community. This approach could stimulate the individual as a free subject and the individual's critical vital power. Since the individual is a dichotomy, it might be possible to form a lifestyle with others.

I consider the writing of this book to be the activity of a 'tradition individual.' In the process, I have become increasingly aware of being part of a community, a society, and above all, a tradition. This greater awareness includes the painful realization that dualism is deep economy's basic structure.

MONEY'S POWERFUL ROLE

Daly and Cobb express sympathy and reservation toward deep ecology. Defining God as the final truth and "omniscient and impartial unifying source of all," they freely locate God in both the biosphere and the Bible. This prevents them from analyzing deep economy's ideological claim of protecting and stimulating the common good. Since the common good is the subject of their book, this oversight is rather unforgivable. Their mistake is that they settle on the conclusion that market economics has an idolatrous nature without examining deeper avenues. Their conclusion is therefore unconvincing.[39]

I agree that economics has a religious nature. But this platform is not sufficient to fight economics' totalitarian claim without a deep analysis of its background.[40] The economy's historical development and its relation to European Enlightenment must be traced in order to know why this totalitarian claim is possible.

We need analysis and theory regarding the 'invention' of money's role as the savior and sustainer of our existence. Where on earth does money come from? Who creates money and how? What kind of communal cooperation is behind this creation process? What is the function of moral and legal rights in the process? These are the fundamental questions.

Daly and Cobb address these questions in the afterword of their book's second edition. Here they confirm our living in an exponential growth culture and our inability to escape this type of culture. They mention Marx's theory of capitalism, including the commodities' exchange value at the heart of the capital-accumulation process, but they do not seriously analyze the *creation* of money. Instead, they say that most money is the creation of commercial banks.[41]

Their awareness of the historical discussion about this issue is evident by their explanation:

> Part of the confusion [about the creation of money] may have been the distinction between money (customary means of payment) and legal tender (money that one is legally obliged to accept in payment). Banks do not create legal tender, only governments can do that. But banks do create customary means of payment. (p. 415)

Regarding the creation of money, Daly and Cobb complain that the discipline of economy provides insufficient reflection. The only well-known economist who thought about money and finance in a radical way was Frederick Soddy, and initially, he was not even a trained economist. Soddy "considered the respected canons of sound banking to be themselves little more than funny money schemes to mystify the public for the enrichment of the bankers and their class" (p. 420). The trick is "to convert wealth that perishes into debt that endures, debt that does not rot, costs nothing to maintain, and brings in perennial interest" (p. 423). Soddy saw this as early as 1926, and it has increased exponentially since then. Is the creation of money one big mystifying trick?

Daly and Cobb cite Soddy extensively to express their indignation toward the banks' financial behavior and power. I agree with their informed critique and find their descriptions quite apt and enjoyable: "the idea that all people can live off the interest of their mutual indebtedness is just another perpetual motion scheme—a vulgar delusion on a grand scale." And after a complicated technical argument, Daly and Cobb bring us to their conclusion:

> The real lender is the community, which ends up holding more money-debt and fewer real assets. In other words, the community has abstained from the use of real assets, making them available to the bank's borrower in exchange for the money created by the bank and loaned to the borrower. . . . Prices are eventually bid up, since *ex nihilo* creation of money (demand) is easier and faster than *ex materia* creation of new physical wealth (supply). The very existence of the bulk of our money now depends on this debt never being retired, only continuously rolled over. . . .

A first step away from a culture of exponential growth, and toward a culture capable of dealing with problems of non-growth, would be to restrict the ability of money to do some of the things that wealth cannot do. This seems to mean two things: first, limiting the indefinite exponential growth of money values implicit in projections of compound interest growth over long periods; second, limiting the 'conjurer's trick' of creating money *ex nihilo* and then destroying it. That power would be taken away from the private banks, and reserved to the governments. (pp. 426–8)

What Daly and Cobb ultimately propose, borrowing from Soddy, is a different management of money. Abandoning all conjurer's tricks, politics should take a firmer grip on the process of money creation. The banks are evidently the conjurers. Government would handle money and wealth more realistically.

Daly and Cobb's analyses and proposals for improvement sound healthy. They confirm my long-held suspicion that there is much conjuring in the current money system. Demystifying the banks' profitable games and reallocating power might offer real improvement. Yet again, their economic reasoning does not go deep enough. Daly and Cobb's discussion remains on the level of Adam Smith's money theory, referred to in the first chapter as the Trinitarian formula. The banks succeeded in expanding money's three original sources to four, five, or even six. The source that precedes wages, profits, rents, and the rest is still not included in the discussion. This fact, however, only proves that this root source is a feat of brilliant conjuring. Marx calls it the commodity's fetish-character, which he considers the true mystery.[42] We could say that all of Marx's economic work is dedicated to the revelation of this mystery.

The mystery is, how was the current currency—money—originally created? To use religious terms, as Marx did, we can ask: Where should we locate the first step in the process of transubstantiation? Regardless of whether one agrees with Marx, deep economy thinkers should attend to this fundamental question. The original abstract conjuring was a very concrete action; indeed, it was the transformation of nature into money (the transformation of natural, intrinsic value into monetary value). The only way (and this is the critical thinker's method) to unveil the mystery of changing natural value into monetary value is to realize that the process is enacted in two phases: the phase of commodity's production, and the phase of commodity's exchange. If we understand money's value as surplus value, we can define both phases as production and realization of profit. Why profit? Because in the (free) market system, money no longer has a neutral function as the means of 'making trade easier.' This is a hard fact, in spite of what the economic textbooks tell us.

To summarize an involved process, our competition-oriented society deems that commodities and services should be as cheap as possible, but they still have to yield some profit on the market. The only way to achieve this is exploitation, particularly in the initial phase of moneymaking or creation. 'Exploitation' is a loaded word, but there is no better expression for the way in which our society uses raw material, energy, and labor. As previously stated, 'exploitation' includes 'exclusion.' Only promising humans and parts of nature can be exploited. 'Promising' means that there is a reasonable guarantee of profit.

Daly and Cobb discuss money mainly as far as it circulates in wealthy countries. In the global environmental discussion, however, we should never forget the gulf between the so-called developing countries and the developed countries. The conjuring-like appearance and disappearance of money, in which private banks play a central role, stimulate wealth in rich countries. This is done by exploiting poor countries, including (their) nature. The exploitation-exclusion mechanism, which includes parts of the world in exploitation and then excludes them from wealth, causes money's concentration in certain parts of the world. The real conjurer's trick is the transubstantiation of Third World natural richness into First World money. It seems that money is capable of serving as a contributor to capital accumulation only in places where money already exists, where private banks have the trust and license to create the money.

Before we can go further with a deep discussion of money's creation and flow, we must settle other issues. We must begin by recognizing that social ecology is affiliated with deep ecology. The social ecologist Murray Bookchin reveals an ongoing controversy between the two interdisciplinary fields:

> If the deep ecology principle of 'biocentricism' teaches that human beings are no different from lemmings in terms of their 'intrinsic worth' and the moral consideration we owe them, and if human beings are viewed as being subject to 'natural laws' in just the same way as any other species, then these 'extreme' statements (e.g. Famine and AIDS are nature's revenge for over-population and ecological destruction) are really the *logical* conclusion of deep ecology philosophy.[43]

Here we meet the same objection that Daly and Cobb formulated concerning Naess and Sessions's eight principles: that higher forms of life are more valuable than lower forms. Yet here we must ask an intriguing question of biocentrism advocates: Do they argue why their community concept is superior for human beings? And can Daly and Cobb uphold their arguments while working in critical solidarity with deep ecologists? Bookchin is not only critical of biocentrism but of interrelatedness as well. His departure point is power. In *The Ecology of Freedom*, he concentrates on the various forms of hierarchy and domination that are present in all known cultures. The scholar Joseph Des Jardins comments on Bookchin as follows:

Bookchin suggests that social structures of domination preceded the domination of nature. . . . Societies characterized by a high degree of hierarchy
are also likely to abuse and damage their natural environment. Social
hierarchies provide both the psychological and material conditions, the
motivation and the means, for exploiting and dominating nature . . . success will be understood in terms of dominance and control. The more
people who work for you, the greater wealth, power, and status you have.[44]

In my opinion, Bookchin is correct in attributing priority to the human
drive for power, dominance, and control. Disappointingly however, he does
not work out this datum for our phase of social history, where dominance
and control have a special, fixed function in the economic process. These
human drives are conditions for the creation of money. Bookchin misses
this point and gets stuck in his anarchist conception of freedom, a position
he calls the 'organismic tradition.' This tradition is concerned with the dialectical relation of community (organic society) and individuals. Bookchin's
anthropological presupposition is the individual's 'self-determining activity,'
which entails an anarchist's conception of freedom. In philosophical anarchism, there is no place for analysis of a general dominant ideology, such
as the necessity of growth and capital accumulation. All claims of authority
are interpreted as forms of the individual's power or coercion. The only
solution in this approach of social reality is the just community that is meant
to serve common needs and goals. Here, Bookchin and Daly and Cobb are
not far apart.

The situation is much more complicated than these community authors
are willing to admit. It is one thing to fight coercion and domination; it is
quite another to define and establish freedom. In old Athens, freedom was
available only to male citizens. In eighteenth-century Europe, a threefold
freedom was proclaimed: freedom of spirit (autonomy), political freedom
(democracy), and economic freedom (free market). The German Kant, the
Frenchman Rousseau, and the Scotsman Smith all wrote their famous books
on these subjects. In the turbulence of Western history, a certain hierarchy
among these three fields developed. Concrete, materialistic conditions enabled the other forms of freedom. The free-market economy dominated all
areas of life. It offered all people the opportunity to participate in production
and trade, under the condition of obeying the system's laws. Adam Smith
called this the right moral behavior. I state that this is indicative of God the
Father (economics), God the Son (politics), and the Holy Spirit (morals).

Opposing my somewhat cynical depiction is the deep ecologists' conception of a divine power working in natural, biospheric relations. Deep ecologists, especially social ecologists, pretend that their God is able to defeat
the economic God. Apparently, the eco-divine power is a revelation of the
true God, whereas in the realm of economics, we are dealing with

pseudodivinity. A good question is: How do they know? After all, doesn't the economic God have the advantage? The freedom that the economic God promises is what modern people are striving for. It is the new freedom, acquired via a lengthy process of secularization and technological and scientific development. Natural forces are under the control and domination of the human species. If we want a real confrontation, which is not taking place in deep ecology or in social ecology, we must look at modern economy's essence and its consequences.

NOTES

1. Vincent Vycinas's concise translation of Heidegger's meaning of dwelling, cited by Bill Devall and George Sessions, *Deep Ecology: Living as if Nature Mattered* (Salt Lake City: Peregrine Smith Books, 1985), pp. 98–9.
2. There is an interesting resemblance with biblical thought. The Hebrew *dabar* is generally translated in the Greek as *logos*. For instance, in the prologue of the Gospel of John: "In the beginning was the *logos*, and the *logos* was with God, and the *logos* was God." *Dabar/logos* means 'word,' 'critical story,' 'happening that makes sense.' In this case, *dabar/logos* is not opposed to *muthos*, but it criticizes and overcomes the *muthos*'s divine, religious character. For this reason, the *dabar/logos* is often interpreted as Christ himself, the preexistent Son of God.
3. Cf. Steve Chase (ed.), *Defending the Earth: A Dialogue Between Murray Bookchin and Dave Foreman* (Boston: South End Press, 1991).
4. See Martin P. Nilsson, *Geschichte der Griechischen Religion*, Vol. 1, *Die Religion Griechenlands bis auf die griechische Weltherrschaft* (München: C. H. Beck'sche Verlagsbuchhandlung, 1967), p. 738.
5. See Charles Taylor, *Sources of the Self: The Making of the Modern Identity* (Cambridge: Cambridge University Press, 1992).
6. Cf. Karl Löwith, *Meaning in History* (Chicago: University of Chicago Press, 1949), and Erich Voegelin, *The New Science of Politics. An Introduction* (1952; Chicago: University of Chicago Press, 1987).
7. See Dietrich Bonhoeffer, *Letters and Papers from Prison* (New York: Macmillan, 1971), especially the letter dated June 8, 1944. I agree with his positive approach and serious questioning of the adult (emanicipated, enlightened) European culture: "The attack by Christian apologetic on the adulthood of the world I consider to be in the first place pointless, in the second place ignoble, and in the third place unchristian. . . . The question is Christ and the world that has come of age" (p. 327).
8. George Bradford, *How Deep Is Deep Ecology? With an Essay Review on Woman's Freedom* (Novato, CA: Times Change Press, 1989).
9. Herman E. Daly and John B. Cobb Jr., *For the Common Good: Redirecting the Economy Toward Community, the Environment, and a Sustainable Future* (1989; Boston: Beacon Press, 1994).
10. Bradford, *How Deep Is Deep Ecology?* p. 6.
11. Ibid., p. 10.
12. Chase, *Defending the Earth.*

13. Devall and Sessions, *Deep Ecology*, p. 70.
14. Ibid.
15. Ibid., p. 69.
16. Daly and Cobb, *For the Common Good*, p. 384; emphasis added.
17. Ibid.
18. Here the "I" and the "other" are sharply divided. Acknowledging this human practice, Levinas highlights the other's power, invading the I's well-protected center. See Emmanuel Levinas, *Totality and Infinity* (Pittsburgh, PA: Duquesne University Press, 1969).
19. Bradford, *How Deep Is Deep Ecology?* p. 13.
20. Ibid., p. 22. The critique is found within the context of a comparison with Malthus's scientific reductionism.
21. See chapter 1 for a discussion of Adam Smith.
22. Daly and Cobb, *For the Common Good*, p. 25.
23. Daly and Cobb quote Nicholas Georgescu-Roegen: "It is beyond dispute that the sin of standard economics is the fallacy of misplaced concreteness" (*For the Common Good*, p. 36).
24. Daly and Cobb, *For the Common Good*, p. 37.
25. Cf. John Rawls, *A Theory of Justice* (Oxford: Oxford University Press, 1971). He presents the community of people in an original position as an 'experiment of thought,' but the line of argument doesn't differ very much from Daly and Cobb's.
26. Recalling the Nazis' use of the term 'community' (*Gemeinschaft*), Daly and Cobb correctly warn against a too naive use of the word. See Daly and Cobb, *For the Common Good*, p. 170.
27. See chapter 1.
28. See chapter 1.
29. Daly and Cobb, *For the Common Good*, p. 139.
30. See two recent publications: John Rawls, *Political Liberalism* (New York: Columbia University Press, 1993), and Joshua Mitchell, *Not by Reason Alone: Religion, History, and Identity in Early Modern Political Thought* (Chicago: University of Chicago Press, 1993). Both publications discuss the theories of Thomas Hobbes, John Locke, and Jean Jaques Rousseau on the subject of the rise of the modern 'citizen' and the 'social contract.'
31. Thomas Hobbes still had problems with political equality in the seventeenth century, and he advocated, partly on biblical grounds, a strong leadership. See Mitchell, *Not by Reason Alone*, pp. 46–72.
32. The 'making of the modern identity' (the subtitle of Taylor's book *Sources of the Self*) is closely related to humans being *homo economicus*.
33. Ferdinand Tönnies, *Community and Society (Gemeinschaft und Gesellschaft)*, with an introduction by John Samples (New Brunswick, NJ: Transaction Publishers, 1993).
34. Daly and Cobb, *For the Common Good*, p. 172.
35. Daly and Cobb indicate that Protestantism focused on the relation of God and the human soul and that it lost the notion of the human community as an instrument of grace.
36. Here, the authors refer to Lovelock's Gaia hypothesis.
37. Cf. Taylor, *Sources of the Self*, p. 83, on Kant's stress on freedom as self-determination.
38. Cf. Daly and Cobb, *For the Common Good*, introduction.

39. Ibid., pp. 389–93.
40. Some liberation theologians do the same. See, for example, Pablo Richard, "Biblical Theology of Confrontation with Idols," in *The Idols of Death and the God of Life*, edited by Pablo Richard, et al. (New York: Maryknoll, 1983), pp. 3–25.
41. Daly and Cobb, *For the Common Good*, pp. 410, 414.
42. Karl Marx, *Das Kapital: Kritik der Politischen Ökonomie* (1867; Berlin: Dietz Verlag, 1973), vol. 1, I.1.4.
43. Murray Bookchin, *Defending the Earth*, p. 125, cited in Joseph R. Des Jardins, *Environmental Ethics: An Introduction to Environmental Philosophy* (Belmont, CA: Wadsworth, 1993), p. 248.
44. Des Jardins, *Environmental Ethics*, p. 244.

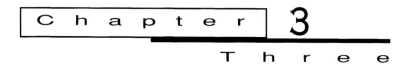

CREATION, NATURE, AND DUALISM

THIS CHAPTER PRESENTS A historical and semantic overview of dualism's development and significance. The concept of dualism bridged the conflicting Hebrew and Greek worldviews, allowing for the development of the Christian worldview. Thus, Western dualism is the pillar of modernity. The central feature of Western dualism is humanity's initial break from nature and its reconnection on the foundation of economics. Without this initial break, science and technology would never have developed. Moreover, Western dualism hinders the reestablishment of global consciousness or cosmic connectedness, particularly in the realm of practical consequences, which concerns economics. We need a new paradigm for our relationship with nature that includes interhuman social and economic relations.[1] In order to build this new paradigm, we must understand the roots and origins of the present and past situations. We must first analyze the origin and development of early Christian dualism. This entails a wide historical and textual navigation through the 'creation text' and ideologies of Greco-Roman, Judeo-Christian, and modern Western history.[2]

Historian Lynn White Jr.'s theory has become very influential in the Western environmental discussion. White asserts that Western society's dominance over nature originated in the creation beliefs of the Judeo-Christian tradition. White specifically locates this origin in the textual story: "Finally, God had created Adam and, in an afterthought, Eve to keep man from being lonely.... And, although man's body is made of clay, he is not simply part of nature: he is made in God's image."[3]

From this Jewish story, Christianity evolved into the most anthropocentric religion in world history. It also led to the conception of time as linear rather than cyclical. Both Greco-Roman and ancient Eastern cultures had cyclical notions of time, so they did not conceive of creation as having a specific point of origin or beginning. In contrast, Christianity has a beginning point and hence an ending point toward which history steadily progresses. In the Jewish creation story, the human is separated from nature, in that the human is more than nature. Christianity evolved the notion that it is God's will for humans to exploit nature for their own purposes. Since humans are 'higher' than nature, they are free to disregard the position and 'rights' of animals and natural objects. The Christian belief in progress(ion), the freedom to exploit the environment, and the grand-scale spread of Christianity culturally and geographically led to the twentieth century's ecological crisis. In terms of practical consequences, scientific and technological advancements have given us the means to destroy our environment.

White makes an important historical contribution when he establishes that Western scientific and technological supremacy began in roughly the thirteenth century, long before the Scientific Revolution of the seventeenth century and the Industrial Revolution of the eighteenth century. White traces this development to regional theological practices and pursuits. In the Greek East, nature was studied as a symbolic way to understand God's communication with humans (natural phenomena as signs). However, in the Latin West, nature was studied as a way to know God's Spirit (natural phenomena examined in terms of elements, operations, natural law, and so forth). And the West flourished:

> By the late 13th century Europe had seized global scientific leadership from the faltering hands of Islam. It would be as absurd to deny the profound originality of Newton, Galileo, or Copernicus as to deny that of the 14th century scholastic scientists like Buridan or Oresme on whose work they built.[4]

Ultimately, these divergent religious pursuits and scientific practices led to Western supremacy and Eastern subordination.

Western religion and science provoked the catastrophic separation of humanity and nature. White cites the decisive break as the moment that science viewed nature as an object. Yet this crucial break did not become an environmental threat until the counion of science and technology in the nineteenth century. This counion, initiated in the interest of human progress and empowerment, resulted in science and technology's overpowering of the natural environment. White places the blame squarely on Christianity:

> Certainly the forms of our thinking and language have largely ceased to be Christian, but to my eye *the substance often remains amazingly akin to the past*. Our daily habits of action, for example, are dominated by an implicit faith in perpetual progress which was unknown either to Greco-Roman antiquity or to the Orient. It is rooted in, and is indefensible apart from, Judeo-Christian teleology. The fact that Communists share it merely helps to show what can be demonstrated on many other grounds: *that Marxism, like Islam, is a Judeo-Christian heresy*. We continue today to live, as we have lived for about 1700 years, very largely in a context of Christian axioms.[5]

White reasons that since our ecological problems are physically caused by science and technology, we cannot look to them for solutions. Instead, he states that we must acknowledge Christian belief and ideology as the root cause and look toward a reinterpretation of religion for practical answers: "Since the roots of our trouble are so largely religious, the remedy must also be essentially religious, whether we call it that or not."[6] White asserts that we should not look toward Zen Buddhism but should stay within our own tradition and practice the humility of Francis of Assisi.[7] At this point, White's interesting and informative historical analysis falls apart.

There are three central problems with White's 'religious remedy.' First, White totally ignores economics. He does not identify the rupture in history that occurred with the counion he talks so much about. In the early Enlightenment, the West was in pursuit of spirituality and truth. By its close, the West had chosen an economic path. A real solution to today's state of crisis must analyze this rupture and, obviously, the field of economics. Second, White incorrectly separates Christianity from other cultures. Christianity did not just grow out of other cultures; it incorporated and retained features of other cultures. So we cannot limit our focus to the Judeo-Christian tradition to find the metaphysical origin and solution to our current problems. Third, White argues that Western dominance originated in a Jewish story. This blames Judaism for creating Western dominance and its consequences.[8]

The fact is, the original Jewish interpretation of the story is completely different from the Christian interpretation, and the same is true for the Jewish culture's interpretation of nature. The story does not form the culture; the culture forms the story through its interpretation of it. A new interpretation will not form a new culture, although it may indicate that a new culture has formed. White's location of Western dominance as originating in a story— a creation story that already had an interpretative history before Christianity adapted it—is a fictional theory that is useless in terms of the current real-world environmental crisis.

In our effort to build a new paradigm of humanity's relationship with nature, we must trace cultures and texts that have affected Western Christianity. Cultural and textual reinterpretation is needed, because modern interpretations have distorted original meanings. We need to acknowledge original meanings as fully as possible. The concentration on Western Christianity is due to the fact that it eventually led to modernity as we know it today. We need to trace the origins and evolutions of interpretations in order to build the new paradigm, the success of which is essential for our environmental well-being.

Humanity and Nature: The Biblical Line

The traditional view that Christianity is the fulfillment of Judaism, and hence is the better representation of God's covenant with humanity, is responsible for the single-origin view of Christianity. The concept of Jesus-as-conqueror presupposes a dramatic division between converts and their former cultural traditions. This traditional view of superiority engendered the notion that Christianity *had* to be spread in the Greco-Roman world. The implication of this triumphant attitude is that the Judeo-Christian tradition influenced people of another culture without being touched by that culture in return. Obviously, such a transaction is impossible. The early dissemination of the Judeo-Christian message throughout the Hellenistic world resulted in Hellenistic culture's influencing Christianity.[9] The dualism that features human domination over nature is a result of the mixture of the Greek and the Judeo-Christian traditions. Thus, the term 'Judeo-Christian' is problematic, because it neglects Greek thinking, which had a profound influence on Christianity.

Until recently, biblical interpretation was Christian culture's guiding principle. Even in modern times, economic prosperity has been explained as God's blessing and election. Of course, the attempt to place original texts within their own contexts poses semantic problems. I address these problems by approaching the Bible as a collection of stories with their own linguistic and historical contexts. In this way, I try to uncover the Hebraic understanding of humanity's relation to nature.

In ancient Israel, relationships were not influenced by the existence of a natural and metaphysical order. Instead, the God of Israel made himself known through a historical act of redemption. This redemption fundamentally negated divine powers grounded in natural religion: God reveals himself in and through nature. The prophetic message entails this critical attitude.[10] Symbols with reference to the sacredness of natural data became insignificant. Instead of being used as signs of everlasting divine power, these symbols were reinterpreted as offices or tokens of the liberating God. They were historically recorded and thus became open to critique. The heretofore assumed divine power of natural order, in which human symbols and offices are representatives, fell under scrutiny because of its inherent ideological character. Righteousness and the proper functioning of God's creatures became the ultimate standards. The 'God' Israel refers to is the God who revealed himself in actions of liberation. Where did this concept of God originate? The telling areas of sexuality and politics should clarify the importance of the question and its relation to the human community and worldview, both past and present.[11]

SEXUALITY

Israel desecrated sacred pagan poles and stones and erased their symbolic reference to fertility. The phallus was leveled, and in its place the Hebrews erected stones that marked the locations where people experienced Yahweh's power and guidance. These symbolic stones were related to stories about God's history with his people.[12] These stories told *pars pro toto* (part representing the whole) of God's liberating work in Israel's days of oppression. We can see that the original, phallic meaning of erected stones was transformed. Israel sharply criticized male sexual power as the origin of fertility. Fertility was important in Israel, but it was considered to be God's gift. Fertility was not directly connected to nature, nor did it carry the dominating feature of pagan fertility conceptions (gift of nature). As the annunciations of birth stories indicate (Isaac, Samuel, Jesus), male potency was of secondary importance. Moreover, the connection between fertility and a divine natural power was abandoned. This is all quite surprising for a patriarchal society.

POLITICS

Kingship was constantly at stake in the ancient world. In most nations, the king claimed the 'natural' right, and he symbolized nature's strength. There could be no dictatorial power in Israel, because the real, messianic king was God himself, who, through his actions, became 'the partisan of the poor.'[13] On a few occasions, Hebrew prophets awarded the messianic title to non-Israeli kings, such as Cyrus, King of the Persians.[14] The messianic king represents the God who liberates his people from slavery. If the king fails to rule in accordance with this God's aims, he comes under prophetic critique and often falls from power. There is no presumption of everlasting kingship. The prophetic message frequently indicates the end of a reign or trouble within the reign. The stories of Saul, David, and Solomon exemplify this. Thus, the Hebrew view opposes the ancient traditional view that the king himself is divine. A traditional example of personal divinity is Egypt's pharaohs and dynasties.

Israel emerged as an antisymbol by pointing at a historical development that opposed the ruling structures. Israel severed the apparent coherence between natural violence and violence among people. Israel claimed that human relationships were not governed by a violent ordering power but by a hidden God, an unknown factor, a Lord other than the well-known lords. This claim for a new Lord was made by the lowest in the social order, the people without a chance, the Hebrew slaves. As Israel's culture and religion grew, it constantly upset the sanctity of Egypt, Assyria, and Babel, the world empires.[15]

In biblical stories of temporally and locally definable antimovements, nature plays an undeniable role as a created reality. Israel interpreted natural powers as providing solid, vital strengths and proofs. In various ways, nature points to God's liberating revelation. For example, creation is praised both as a sign of God's greatness and for the special place humanity has in the world. The creation motive in Psalms, particularly Psalm 8, illustrates this. In addition, many 'miracle stories,' including those of Jesus in the Old Testament tradition, can be interpreted as referring to the role of nature in human liberation. The Red Sea's parting or Jesus' walking on water can be interpreted as a symbolic human breakthrough in the chaotic natural power of water. Other nature miracles show God working through natural means for human liberation: the multiplication of the loaves and stories of healing and resurrection.

All these interpretations of physical reality demonstrate that Israel developed a different vision of nature and creation from its predecessors and contemporaries. Biblical stories and symbols all uphold this position. Under God's command, the sun stands still, it rains in the desert, a donkey speaks, an infertile womb conceives, a virgin gets pregnant, and death must relinquish its hold—all for His nation, Israel (*pars pro toto*, for all humankind). Israel respects nature's greatness, but this greatness is clearly connected with God's liberating action. Jesus summarized God's power over and work through nature when he spoke of the everyday sorrows of food, clothes, and shelter: "Seek ye first the kingdom of God and his righteousness; and all these things will be added unto you."[16] Apply this to society at large, and it becomes clear how a just economy grounded in liberation ethics works.

The mission of managing and commanding the earth is given to humans in the first creation story.[17] This is why the creation story is a focal point in environmental discussions. Genesis states that both humans and nature qualify as creation. Creation was brought forth by a creative power, not by a general natural power such as the *logos spermatikos* (the seed-carrying word), which is said to penetrate everything. If the creative power were such, a godly general ordering principle would force its will onto human acting, analogous to the animal instinct. In contrast to Aristotelian and Stoical worldviews, Israel's creative power is explicitly *dabar YHWH* (Word of the Lord). This means that both humans and nature reflect the God who revealed himself in history as the liberator of an enslaved people. But because this God is creator of heaven and earth, neither humans nor nature can have godly pretensions. Creation itself signifies the process-like historical symbiosis of nature and humanity.

Human rule over nature, derived from Genesis, does not mean that humans are appointed rulers over everything. The first story of creation shows that nature does not rule over humanity and that human power is neither derived from nor legitimized by nature. The human who knows his or her position in the history of liberation must control creation: Humans must make sure that nature does not accumulate more importance than was given to it.[18]

Thus nature was suddenly allotted a human-subjective character. By putting it this way, I avoid metaphysical language in order to sketch nature within its proportions. In the name of just relations, the creation order may not be a symbolic reference to the actual, intended order in political and institutional fields. On the contrary, it can only be a reference to the freedom-establishing God who wants to end oppressive relations. This way, humans have an important place in creation. Just humans (the *tsaddikim*) live in harmony with creation, giving the animals their names (Adam, Genesis 2) and saving creation from destruction (Noah, Genesis 6–9). In the Promised Land, the Israelites live as strangers, as tenants on the outskirts; exploiting their fellow humans and nature is strictly forbidden (Exodus 20, Leviticus 25, Deuteronomy 5). Thus, biblically, humans and nature are not divided in an absolute way; there is no fundamental division between upper and lower people, nor between rulers and the ruled. Dualism is not a feature of biblical thinking.

Naturally, this weakens the thesis that Western dominance originated in the creation story, although Christianity eventually did work the way White describes it. However, this does not mean that the historical functioning of Christianity is derived from biblical data. There are alternative readings of crucial points concerning the relationship of humans and nature in biblical stories. Of course, this presumes that it is possible to return to basic biblical data in which 'creation' is understood as the history of people in their social and natural context within a given period of time and space. These basic biblical data have a narrative character—giving hope in a situation of exile—and do not stand in the metaphysical frame.

There is a profound difference between what is written in the old texts and the actual working of these texts in sociohistory. If we can trace the history of interpretation, we can prevent ourselves from feeling obliged to leave our own culture—including its roots—in order to find a solution, thus meeting the claim that White formulates at the end of his article. The objective is to find and overcome the cultural discrepancies that are rooted in stringent dualism. We cannot redirect the aggressive course of Western history, but we can criticize it on well-founded grounds. And we can propose evaluation and different ways of acting; there is even the potential for rebirth within the schemes of life and death.

Humanity and Nature: The Greek Line

In terms of the contextual character of Christianity's development in Europe, it was an originally Jewish worldview that spread through the Greco-Roman world. During its infancy, the mixing of contemporary Greek philosophy and Eastern mystery-religions produced a dualism between the physical world and the spiritual world.[19] Infant Christianity was very receptive to both the mixing of traditions and the resulting dualism. The basic difference between these two traditions helps explain the West's unique aggressive contempt toward life, nature, and the world.

We already know that Aristotle strived for a life 'by nature' (*phusei*), meaning that the social order and the establishment of the city-state (polis) had to be in harmony with the natural order. The highest form of existence was *eudaimonia*, and everything else was hierarchically derived from it. Or conversely, all things served as the foundation of this highest goal. The polis's economy was based on slavery. Aristotle never hesitated to proclaim that the slave was naturally destined to spend his life in subjection. Nature was the ultimate ruler. Certain positions and developments were judged as unnatural or natural, according to their consistency with the innate order and coherence of things. That was how nature wanted it, and therefore this order needed to be respected and protected. There were upper and lower people; it was that simple.[20]

It is logical to ask how a great thinker like Aristotle could define 'nature' so easily. But it seems to have been a matter of course. The Greek philosophers studied the natural, physical (*phusis*) reality as the final and definite reality. Political, legal, religious, economic, and behavioral rules were all a part of the natural totality. It constituted a cosmic entirety in which, according to Plato, the imaginary world (the reality of ideas) also had a place. Eventually, the *nomos* (law, regularity) depended on the *phusis*, even though they were opposites.[21]

This era of the great Greek philosophers of the late fifth and fourth centuries B.C. is known as the first Western wave of enlightenment.[22] Independent human thinking was emancipated from dependence on mythical and religious thinking; logic (the *logos*) was freed from the mythical (the *mythos*). Myths told of the sense and nonsense of being, about fate and recklessness, about the way humanity stood in reality, and about the misfortune that struck when people resisted the eternal laws. Logical thinking tried to understand the world of gods and half gods, of heroes and normal mortals, as symbolic, natural truths.

This does not mean that the Greek thinkers tried to 'think themselves away' from nature or wanted to position themselves at a critical distance from nature, as occurs in modern times. The Greek focus was to recognize humans as part of nature, to be able to live in a good way, following Plato's idea of the good. The unpredictability of the gods or fate supposedly could be controlled in this way. The mischief of the antinatural life could be shown in a logical way. Hence, all discussions of responsibility regarding right government, warfare, money, slaves, friendship, and virtue could be explained by understanding the natural order.

We can conclude that nature did have a divine character in Greco-Roman philosophy. The mythical shapes of gods did not disappear as a whole; their ways of being were translated into terms of reason and nature.[23] Human thinking and understanding held a central position; in fact, rationality received an exalted, divine character. Yet it was never separated from the natural conditions with which humans as well as other creatures had to live.

The Greco-Roman Stoical school developed the doctrine of the fertile, seed-carrying World-Reason. This *logos spermatikos* can be regarded the (Holy) Spirit, the World Spirit. It was a godly gift of nature, and those who received it were blessed. Thinking, fed by the fertile World-Reason, led to harmony and unity. Just as in nature, where life force and vitality united the world as one harmonious entity, the cosmic *logos* as World-Reason brought together the people of the world.[24]

In Greece, as well as in Israel, the good society was important. These societies did not have the dualism of a godly and earthly realm, as is featured in later Christianity. The divine, the transcendent, was always connected with reality under the idea of the good supporting the good, an ethical criterium. In Aristotle's thinking, heavenly felicity (the *eudaimonia*) was the eventual realization of the polis's right government. Plato distinguished a visible and an imaginary world, but he did not separate them. There was a dialectical connection between things and ideas, but not an absolute division.[25] The separation of fields such as heaven and earth or spirit and nature took place much later. It is only after this separation that the relation between life and death becomes problematic. We are dealing with the absurd fallout of this separation today.[26]

The Greek and Hebrew cultures espoused worldviews that optimally guaranteed a life justified in the natural reality. The Greek vision took reason as its starting point. The free philosopher explained how we must live as a part of nature. Knowledge of this was viewed as a natural gift in and of itself. A philosopher, such as Socrates, helps people think; like a midwife (*maieutic*), he was occupied with the process of the birth of reason, continuously and with different people.[27] In principle, everyone was involved, although there was some resistance to the development of free reason. Yet most Greeks saw abuse of power and other conflicts as evidence of a failure to live in accordance with nature and the natural principles of life. To avoid these political and social problems, they reasoned that they had to classify things as *phusei* (natural) and *para phusin* (unnatural).

Israel's reasoning began under conditions of slavery, injustice, and male power claims. Hermeneutically, we are entering a tricky field, because 'reasoning' is typical for Greek philosophy. We must keep in mind that it has a different meaning in Israel's narrative culture. The God of Israel opposed the 'natural order,' which was supported by Greek and other ancient cultures. He freed his people from slavery by breaking through the natural order as espoused by the Egyptian culture. In Hebrew thought, the purpose of life could not be explained by natural ideologies. It could be accounted for only by the historical narrative of a liberating God. Israel taught the world to be suspicious of 'living in accordance with nature.' Indeed, it usually means 'living in obedience to the powers that be.' In the biblical narratives, subtle teamwork with nature helps establish a just society. It is perfectly acceptable to suppose that Moses, Elijah, and Jesus used natural energies, but they did so to break through the oppressive character of the existing natural order. The miracle tales are mythologically wrapped examples of this.

We can also refer to the Hebrew reversal of male potency in the contexts of both sexuality and politics. The symbolic figure of the messianic king became central. In him, God's people were represented and personified in the broadest sense: This was how humanity was meant to be, this was creation.[28] This liberating power created a new vision or point of view toward humans, society, and nature: Look through the eyes of the one below, the one who suffers under the fixed, hierarchical social structure. The position of the slave is the criterion for the difference between Hellas and Israel.

CONTINUOUS DUALISM

Christian culture and its approach to nature originated in the combining of two heritages: Greek nature-thinking and the biblical belief in creation. When the two traditions combined, each lost its individual character. It is therefore impossible to recreate or reestablish the original tradition and society. This utopian conception is merely an unimaginative myth. Nonetheless, the 'working history' of biblical explication (in German, *Wirkungsgeschichte*) in the Christian tradition has become a foundation of Western thinking. What we must now acknowledge, however, is that the present life-threatening situation originated in the combination and assembly of Greek and Hebrew cultural and religious elements.

The dualism between the subjective freedom to exploit the earth and the objective need to protect the earth's resources for life sustenance exemplifies this explicative solidity in Western thinking. The true cause of this dualistic attitude, the one that precedes textual interpretations, is the strained mixing of Greek and Hebrew heritage. Indeed, in order to unite their essentially autonomous worldviews, there was a split in the ideologies espoused by each cultural tradition. I am taking a different stand from those who directly connect the Judeo-Christian tradition and the dualistic, imperialistic worldview. The cultures and worldviews of Greece and Israel were connected to small countries and clearly organized communities. The structure of these communities created a collectiveness that is incomparable to present social structures.[29]

When Aristotle wrote about ethics and politics, his concern was a neat polis. Unbeknownst to him, he was on the threshold of a new era. His own pupil, Alexander the Great, would soon become a conqueror.[30] In the succeeding centuries, Greek, Eastern, and Roman cultures mixed, and the philosophy of the neat polis was applied on an imperialist scale. When Greece became part of the Roman Empire, it disappeared as an independent country with its own philosophical and theological contributions.

Israel followed a different path. Christianity as a 'Jewish sect' spread throughout the world and ultimately Christianized the entire Roman Empire, although the Jews maintained their specific traditions. However, at an early stage, we see the problems of a Greek translation of the Jewish Gospel. Language is a symbolic expression of a culture with its own specific worldview. Thus, problems arose for New Testament authors such as Paul and John. How can central motives be translated so that the translation covers the content? Behind this language and translation problem lies the semantic question of whether different connotations can actually be confronted, and whether mutual understanding is possible at all. A famous example is the term *logos*. Is this the correct translation for the Hebrew word *dabar*? What does John actually mean when he says that the *logos* has become incarnate (John 1:1)? Does he mean that the Greek natural World-Reason has become incarnate? Or is he referring to the creation act of the God of Israel?

Obviously the Greek *logos* understanding prevailed, but it lost its connection with the *cosmos* and the *phusis*. To 'save' the God of Israel, an early dualism was introduced, and it never completely left Christianity. This dualistic conception was the only way to unite the Greek (cosmic, static) and the biblical (historical, dynamic) worldviews (albeit this is a reductive description). But what is this 'thinking in two spaces'?[31] It has been described in many ways: heaven and earth, spiritual and material reality, church and world, spiritual empire and worldly empire (often seen as the two different fields of Christ's mastery). This division extends into the temporal as opposed to the eternal, the time before and after Christ, the beginning of the 'Christian' era.

Neither the Greek nor the Jewish tradition can be regarded as dualistic in terms of a division between humans and nature. To the Greeks, nature itself was imbued with godly 'meaning.' Of course, felicity was often seen as the empire of the gods, and the Platonic empire of ideas was all too often drawn as a reflection of human reality. At the same time, though, reason lies in the natural order, along with the pure identification of the natural right of ideas, as the idea of justice shows. Stoic thinking presents it implicitly: the principle of imperturbability in the penetration of the universe and the cosmos with godly reason.

The Greeks never separated the natural and the spiritual in such a way that created a true dualism between humans and nature. Although in many cases humans failed to live by and to regard natural laws (*phusei*), the ambition to improve was always the ambition of a life closer to the natural order. This allowed for distinctions among people (barbarians, slaves, women, and free male citizens) to remain, but property could only be relative. Abundant moneymaking was wrong because it was unnatural. The metaphysical legitimacy of social relations was in nature itself.

Similarly, Judaism is not dualistic, because God connected himself to the history of humankind rather than to nature. Natural context plays an important role in the forming of alliances, but the knowledge of the godly word (*dabar*) lies in historical events, which were kept in memory by storytelling and erecting symbolic signs.

Regarding these two elements, the 'solution' was found in the stringent separation of nature from God and humanity. This evolved into a division between humans and nature in modern times, and many elements were lost in the chasm. For example, Greek cosmic notions have lost their strength. And what of the character of Israel's God, who, being concealed (*ex nihilo*), reveals himself in miraculous histories of liberation? Although these stories go against the normal or 'natural' course of events, they do not take place as a salvation-history in a separate, spiritual space. Of course, this has changed over the course of history, and the term 'dualism' has acquired additional meanings in our secularized culture.

THE ATTRACTION OF DUALISM

The dualistic experience of reality has been an attractive perspective for ages. Dualism offers the possibility of connecting irreconcilable qualities. It allows for the inner connection of seemingly contradictory things. The attractiveness of dualism is its potential to unite contrasting aspects of separate cultures and religious systems. The dualistic image of reality developed in stages. In biblical thinking, God is neither immanent in nature nor outside reality. This approach to nature is incomparable with Greek thinking. The God of Israel is concealed; he is revealed only in the history of his people. The Hebrew experience of human freedom and liberation is important, because there are no natural powers, stipulations of fate, or dictates of morality through which God makes himself known. God's rules are laid down by the human representative chosen to help his people live in a free country. This is both a hazardous and an ironic interpretation of revelation—hazardous because power mongers can abuse this interpretation, and ironic because in God's name we laugh about these figures in a liberating-moderating way.[32]

Thus, nature serves as the background for human history. God makes himself known in the other. Humans need a certain connection, both with one another and with nature. The call that emanates from the connection is referred to as 'God.' God, humanity, and nature are 'moving' together. They are 'in process,' they are making history together. This is unthinkable in Greek thought, where nature is static or cyclical. Humans can disengage themselves from nature somewhat through the *logos*, but this *logos* eventually is connected to *phusis* (nature) and *nomos* (law, principle). It is unwise to act and live unnaturally. Plato's ethical criterium, the idea of the good, cannot be unnatural. In trying to unite these two worlds, Greece and Israel, we must resort to dualistic frames of thought. Reality can then be thought of as nature, static, with its own laws. As far as humans are bodies, they too are a part of it. Separation of the soul allows for a combination with God's revealing action.

The forms of dualism are often absurd. After its rise, it seems that there is no way back. Heaven is God's place, earth is humans' and nature's. Humans are part body and part soul, both mortal and immortal. Life is divided into two—the temporal and the eternal. Only humans are destined to eternal life (or to eternal damnation). The rest of nature simply dies. Thus, the human is an uncommon creature: Human origin, destiny, and meaning have a metaphysical nature. What happens here is truly revealing: God makes a covenant with humans; he will not forsake this particular group. But this covenant is possible only when the human composition is split: part temporal and part eternal.

Thus the Old Testament gets an emblematical character: The liberation of God's people is the earthly reflection of heavenly redemption. Hence allegory prospers. The schism solidifies in a poignant way: the spiritual godly human versus nature. Christianity considers itself the heir of Israel, but Israel has had its day and now serves only as a 'prediction of what will be' for the Christian world. The mere existence of Jewish people practicing their own religion incites Christian anger. This anger is the root of Christendom's history of anti-Semitism.

Letting the Old Testament speak for itself, we can see that the underlying concept in God's association with Israel is that they can survive miserable natural conditions. The forty-year desert journey shows how the Israelites applied and utilized natural resources and energies. As the Israelites experienced, a subtle teamwork between human and nature is necessary, the perfection of which may take many years. But a Promised Land awaits in the distance. Years of communal living in an inhospitable desert results in the ability to inhabit the land in a fruitful way. The God of Israel is the God of life. He provides, as the Exodus story tells us, the possibilities for his people to live in *shalom* (peace). But the people must make shalom a social reality. Only then is nature called creation. Natural circumstances are signs of God's on-going liberation. Sustenance appears when the people are ready to give up. This pattern repeats itself again and again.[33] There is no supernatural interference in the sense of magic. Rather, the possibilities of heaven and earth are revealed to people who are receptive to experiencing creation in this way, and in this sense they have 'supernatural' experiences.

The cooperation among humanity, nature, and 'God-the-liberator' is important in all this. It points at a close bond, which is the central notion of covenant. It does not suggest dualism. When manna falls from the sky, the point is the distribution of food and the prohibition of oversupply (Exodus 16). There is no such thing as a separate area where actual life takes place. The thought that the human soul is not bound to time and space, being a 'spiritual' part of a larger transitory natural existence, is an altogether different thing.

Dualism is strongly preserved in the course of Christian history. Obviously, the Hebrew notion of covenant does not fit into Christian ways of thinking. Hence, it is preferable to refrain from the term 'Judeo-Christian tradition.' This term should be reserved for critical discussions in which Christianity needs to be referenced in terms of Jewish philosophy. The term 'tradition' is also problematic. Christianity severed itself from its founding tradition. The expression, if used at all, can relate only to the origin of Christianity.[34] Can we say that Christianity leans more on its Greek origin? Yes, and no. Yes, because great thinkers such as Augustine and Thomas Aquinas go back to Plato and Aristotle. No, because the natural reality takes a different place in Christianity than it does in Greek philosophy.

In Christian thought, anthropocentric dualism gains preponderance in the development of dogmatic belief. For example, the 'new Adam' usurps the 'old Adam's' position on the battlefield of Christian history. In this typology, the new human must repress natural tendencies and passions, since Adam's sin is identified with sex and power early in the Christian tradition. The new human still sins but ultimately overcomes evil through a reconciliation with God. In gnosis, the material equals the wrong. The material is the evil that chains the good human soul. The following defines basic premises of the Christian Gnostic:

1. The highest God—that is, the Spirit—has nothing to do with this world.

2. This world is the work of the lower God we know from the Old Testament.

3. Humans carry in their souls a little piece of the highest God himself.

4. The highest God sends Christ to redeem this piece.

5. Christ's doctrine and example make the spirit disengage itself from materiality and return to the highest God.[35]

Although the early church rejected this doctrine as heresy, the Hellenistic-Roman world accepted this vision. It fell under the spell of this attractive form of dualism. In Western culture's church-dominated history, there are two related developments that originated from this dualistic worldview: the impulse to penetrate natural reality, and a hostility toward nature, which concerns the liberation of the mind.

DUALISM IN A TRIPLE SENSE: A HISTORICAL TRACK

Dualism continues to have a considerable influence in three important areas. First, dualism plays a role in the separation of the worldly and the spiritual realms—the 'two empires doctrine.'[36] Second, it affects subject and object division. Humans are separated from material reality. There is the living subject, and then there is dead, mechanical matter.[37] Third, dualism contributes to the problem of contrarieties in ethics. These three elements are historically intertwined. The 'two empires doctrine' was spread by Martin Luther. It confirmed the contradiction of politics and religion as two different fields. However, this dualistic construction preceded Reformation. When Christianity became the religion of the Roman Empire, wherein the Jewish and Greek cultures were united, the separation had to be installed.

We first see this separation in Saint Augustine (354–430). He distinguishes between the city of God and the earthly city (*civitas dei* and *civitas terrena*). The 'grace-empire' of God's love (*civitas dei*) fights the empire of carnal persuasion (*civitas terrena*). Both empires have a supernatural origin: The former has a heavenly one, and the latter has a diabolic one.[38] Thus the former will end in eternal glory, the latter in eternal damnation. Augustine reads the Bible through neo-Platonic eyes.[39] God's eventual entity, intellect, and will are localized in the *civitas dei*, which is analogous to the Platonic world of ideas. But God's tracks are definable in the *civitas terrena* as well. The earthly state comprises both nature and human society. It does not automatically coincide with the institutionalized state, just as the 'God-state' does not necessarily coincide with the church.

Augustine formulated these ideas in the context of the Christianization of the Roman Empire, meaning that the state is called on to conform to God's eternal will. It concerns a deeply rooted ethical call with clear intentions. The future only appears to belong to the hedonistic culture; its eventual collapse is guaranteed.[40] The people who participate in the spiritual world of ideas are partners in God's act of creation. Hence, Augustine offers a specific conception of humanity with a spiritual origin.[41] We can call this dualistic, because the human's free will is disconnected from physical and social dimensions.

Thomas Aquinas (1224–74) explicitly includes dualism in medieval Christian philosophy. The scheme is now regarded in a natural-supernatural combination.[42] Natural reason is God-given, and its place in the order of creation and society, and hence in reality, is logical. Supernatural reason corresponds with nature. Like Aristotle's *eudaimonia* (felicity) as the ultimate goal of reality, the supernatural leans toward divine order. The closest approximation of happiness is the aim of this order. It concerns the 'cosmic argument of God's existence,' in which God's revelation and 'the natural light of human reason' work together. According to Thomas Aquinas, human virtue is the preeminent field through which God's love expresses itself. The right cultivation of the virtues is reached by natural reason, which gets its impulses from God. The correct functioning within the natural order is a condition. The church is responsible for the consolidation of the order; it is the mediator, the mediating institution between the natural and the supernatural. This is why the Holy Spirit is bound to the church. The Holy Spirit cannot work in the world without church mediation.[43]

In this scheme, unity between natural reason and God's revelation is almost reached through an extreme application of Aristotle's concept of nature, particularly in the divine intention (the teleological tendency) of the natural, functional hierarchy.[44] Yet dualism remains, because there is no cohesion between the recognition of God in natural reason and God's self-revelation in supernatural reason. In fact, they cannot coincide without threatening biblical elements. The Greek variant is not really recognized, because it implies that God can only be known in nature. But neither can the Israel variant be adapted. In this variant, God and humanity are involved in an ongoing, process-like history within which nature plays its own specific role. Therefore, we see a static ontology in Thomas Aquinas. In order to include the Jewish God, there must be a dualistic scheme. Thomas Aquinas seeks God's nature, which is possible only through church mediation, not through history.[45]

Dualism remains intact throughout the Reformation. Luther (1483–1546) speaks of the two regiments or two empires of God,[46] whereas Calvin separates human society from the holy, unattainable God. Humans need to live for God, and it is important that the right conditions be created in a theocratic society. Yet at the same time, the doctrines of predestination and pre-ordination deeply separate humanity by decreeing that society and nature are on one side and God is on the other.

The only way to join the traditions of Athens and Jerusalem in Western Christian society was to employ dualistic patterns of thought and belief. Again, each tradition can hardly be called dualistic in itself, especially in consideration of the distance between humanity and nature. They become dualistic when they are brought together.

But why did humans choose this fatal, dualistic path? One reason is that dualism allowed phenomena to be explained as the work of God's hands. Natural disasters were interpreted as punishment for human sins. This set up the opposition between God and nature. Christianity generally inhibits the sense of being a part of nature and promotes the desire to be on God's side. This 'being on God's side' facilitated the transition to a 'God the Father' market economy. Calvinist predestination and the notion that prosperity was evidence of election arises from this transition.[47] The Reformation stressed the personal salvation of the believer by God's mercy. As a justified sinner, the human lives 'with one leg in heaven.' Thus, humans are far from nature; creation is temporal, the eternal is everlasting. Only the latter is truly important.

Modern humanism was the logical next step in this development. Modern humanism originated during the Reformation. Humanism claimed that humanity had a far more central existence. It pleaded for the relative independence of human reason; consequently, it opposed church mediation between God and humanity. The assertion of autonomy broke the established dualistic pattern of church versus world, chosen versus damned. The emancipation of *homo religiosus* parallels the liberation from natural restrictions and disasters. In the natural arbitrariness, *homo religiosus* saw signs of God's actions. The new autonomous human can control natural arbitrariness through research and development.

The secularization of western Europe during the eighteenth century initially meant a maintenance, and even a growth, of dualism with a new form and content. Humans who localize God at history's inception (for example, the Deist's watchmaker) usurp God's place themselves. God used to be in charge, but now the educated, enlightened human is in power. Heaven, as God's residence, vanishes. It is replaced by the position of the freely acting human. However, in order to protect the new human position, humans need to take a defensive stance toward an unreliable nature.

Knowledge of natural laws makes it possible to control them. Hence, natural laws became the subject of study in physics. The goal was to discover how nature operates. This requires a stringent dualistic concept of reality: The ruler (explorer, producer, subject) is separated from the ruled (explored, produced, object). After the Middle Ages, there was a rapidly widening gap between natural law and natural right. By the eighteenth century, natural law had acquired an ideological character. The natural right to property and appropriation has largely determined our ethics. The recognition and abolition of this deeply rooted dualism is necessary for constructive and effective environmental ethics.

REDEMPTION, LIBERATION, AND DUALISM

The deep tracks of dualism in our culture outline the long history of separation between humanity and nature. We can also discern a connection between the religion-dominated era of the Middle Ages and our modern, structural, atheistic era. This connection is disputed occasionally, on the grounds that a fundamental rupture occurred in Western history that severed it. But although this rupture indeed applies to the market economy and democratic politics, it does not apply to the prevailing worldview and philosophy of nature. God's disappearance as 'working hypothesis' and culture's disregard for God (*etsi deus non daretur*) do not erase the 'Christian Occident' from the face of the earth.[48]

The religious theme of *redemption* exemplifies this. In Israel's tradition, the redemption from Egypt was crucial; it meant freedom from oppression and slavery. In Christianity, however, redemption has been reinterpreted as a spiritual condition. It is easy to understand this reinterpretation when we consider the establishment of dualism: spirit and matter, mercy and nature, soul and body, heaven and earth. Undoubtedly, this type of dualism has become commonplace in the modern age. Secularized redemption features faith in visible progress. It is the result of humanity's ideological triumph over nature's arbitrary ruling. The first step in the emancipation from nature is a dualistic disengagement. Humans surpass the natural conditions by thoroughly researching the laws of nature.[49]

Natural science and technology are new in history, but the attitude from which they stem is very old: Dualism historically entails a contempt of nature. Nature needs to be subdued and conquered. Humanity needs to be liberated from it. Of course we wholeheartedly deny this contempt, but it is demonstrated in everyday practice. Moreover, when we redefine this 'contempt' as 'admiration,' language becomes the means for dichotomous magic. A glance at colonial history says it all: That the Europeans considered their race superior is evident by their reference to Africans and Indians, who were part of 'nature,' as 'savages.' Savages could be enslaved because they had no knowledge of the European's spiritual elevation. The savages could be conquered, subdued, and enslaved because they were a part of nature. And as such, Western culture did not consider their behavior, morals, and religion as qualifying for spiritual elevation and its attendant earthly privileges. Stories of the 'noble savage' and the 'original position' of people confirm this situation.

Looking at the present contradictions among cultures and between humans and nature, we can trace a continuation of this dominant ideological aspect. An elite of Western male citizens considered itself a 'culture supporter': As members of a spiritually distinguished group, they transcended the directly nature-bound others. In other words, this 'unique' human type differed from the human others and from the material other. In terms of modern dichotomies, those who differ are of the other sex, the other race, the other class, the other species. The connection between social power and position was at stake. The socially better-off were allowed to force their will onto the socially weaker. The people in power were free to reshape the other in order to use them for specific purposes that benefitted the power holder. This principle became apparent in economic, cultural, and sexual relationships. It has been enforced both politically and juridically.[50]

Thus, premodern and modern times are connected. In general, this continuity is not supported by the public. Humans still continue to claim a 'heavenly' position, but they do so in a hidden way. Even in their earthly existence, humans are heavenly creatures. The development of modernity concentrates on a universal human type. The Western bourgeois individual stands as a model for all the others and is represented by a certain type: the white (race) male (sex) human (species) who can call himself a free citizen (class). The continuous sociohistorical process belongs in the spotlight of public attention because it can help us formulate a critical environmental ethic.

A problematic consequence of dualism is alienation. Human individuals are alienated from the reality on which they depend. Thus, they seek a new connection to alleviate the alienation. In an attempt to establish control, the free white male citizen lets his fellow creatures and nature work for him. Humanity—that is, the group of people who pretend to represent the human being—empowers itself over reality as an autonomous subject.

There is another element at work here. Without the mediation of money, the entire development is impossible. In terms of Western development, money ironically means both everything and nothing. In a nihilistic society, it means power. The postmodern thinker Gianni Vattimo justly defines nihilism as "the reduction of Being to exchange value."[51] In a nihilistic society, the joint dependence on exchange value and the oppressive relationship between power-holding humans and their fellow creatures become clear.

Thus, the Western type of humanity becomes the ultimate goal for all people. This 'ultimate goal' is an ironic imperative: Power in a nihilistic type of society is absurd but effective. Dualism continues as a dichotomy in humans themselves. The representatives of *homo economicus* have friendly and adamant faces, human and instrumental sides. The emotions that they allow cannot, metaphorically speaking, communicate from one cerebral hemisphere to the other. Whether or not the countenance of the other can break through this schizothymia, reacting to an impulse for self-being and authentic existence, remains an important question.[52] But it is certain that before any breakthrough can occur, the absurd duality of Western humanity must be uncovered and acknowledged.

But for now, everyone who wants to be like him must take part in the competition, materially and spiritually. Competition dictates that some control and others are controlled. In Europe, the latest trend is the politician's plea that we need to make way for the 'calculating citizen.' According to this claim, a decrease in governmental control is needed for the citizen to finally and truly develop in a real free society. This means total 'free enterprise.' This trend is merely a continuing consequence of the entire development. The underlying premise is that independent-acting citizens, concerned with their own business, will naturally develop good characters.

This development accelerates Western society's perpetual dualism. Indeed, it binds us to systems and system laws. The independent civilian tries to implement these laws but often fails. The result is a definite division in society.[53] The 'calculating citizen' who achieves a state of well-being does so through an extreme dichotomous or even dualistic attitude. Everyone must consciously obey the system without scruples. This means that all must participate in the law and the profits of exploitation. 'Exploitation' is used here to mean distracting energy and power from the other, materially and personally, in such way that the other cannot restore its original balance. This implies an unequal relationship between the one that takes (forces) and the one that gives (is forced). This force relationship is a condition for the creation of money. If we use money, which all must do to survive in society, we participate in this relationship of force.

It is important to realize that modern forms of dualism always go with a specific kind of solidarity. Humans distance themselves from nature so that they can connect in a totally new and 'complete' way. Humans have made the transition from a natural connection to an economic connection with nature. And the new economic connection compulsively denies dualism as the basic structure of the Western soul. This compulsive denial blocks the development of enviromental ethics, because the other is considered or even proclaimed part of the powerful subject's self, and thus it is derived from its otherness.[54] In the next chapter, I elaborate on this state of affairs, stressing the system's sacrificing effects as a dominant characteristic. As long as this treatment of nature and other human beings continues, environmental ethics will stay weak and without the effective development that is necessary for our survival.

NOTES

1. The way I use the term 'paradigm' is well defined by Anne Primavesi, *From Apocalypse to Genesis: Ecology, Feminism and Christianity* (Minneapolis: Fortress Press, 1991), p. 18: "'Paradigm' ... can mean simply a standard model or an exemplary occasion. In a religious or philosophical sense it can mean the particular orientation or framework which guides the human spirit authorita-

tively. In its scientific sense it covers the whole constellation of meanings, opinions, values and methods shared by the members of a particular group." After this formal definition, she abruptly proceeds to describe the content of the ecological paradigm, indicating its moral implications, similar to Daly and Cobb (see chapter 2): "In its ecological sense it is a construct of reality which sees across the boundaries between species and strives to give value to each in itself and in relation to the whole and not on the basis of a hierarchical concept of use to a dominant species, humanity. It carries with it a commitment to live accordingly." See also the introduction, where I mention Ian Barbour's recent *Gifford Lectures* and his handling of Thomas S. Kuhn's concept of paradigm and paradigm shifts.

2. In my opinion, both parts of the Bible are one. I presume a clear continuity, without the priority that the New Testament holds in Christianity. It is one culture that is underlined by religious-political unity according to place (Israel) and time (tradition).

3. Lynn White Jr., "The Historical Roots of Our Ecological Crisis," *Science*, 10 March 1967, p. 1205. Republished in Donald VanDeVeer and Christine Pierce (eds.), *Environmental Ethics and Policy Book: Philosophy, Ecology, Economics* (Belmont, CA: Wadsworth, 1994), pp. 45–51. Subsequent references are to the original publication.

4. White, "Historical Roots," p. 1204. White asserts that Islam originally prioritized scientific development. Why Islam didn't continue this path is quite complex. Yet Islamic culture has proved to be as unfriendly toward nature as is Christian culture.

5. Ibid., p. 1205; emphasis added. We find the same qualification of Islam, but dealt with much more extensively, in A. Th. van Leeuwen, *Christianity in Word History: The Meeting of the Faiths of East and West* (London: Edinburgh House Press, 1964), pp. 215–57, 382–96.

6. White, "Historical Roots," p. 1207.

7. White foreshadows the Franciscan and Benedictine debate on nature and environment. The Franciscan position derives a biospheric, deep ecological model from Saint Francis; the Benedictine prefers to consider humanity as a culture creating and sustaining species. See Rene Dubos, "Franciscan Conservation Versus Benedictine Stewardship," chap. 8 in *A God Within* (New York: Charles Scribner, 1972). See also Rosemary Radford Ruether, *Gaia and God: An Ecofeminist Theology of Earth Healing* (San Francisco: HarperCollins, 1992), pp. 184–91. She calls the Franciscan attitude "the ascetic flight from earth."

8. The well-known German author Eugen Drewermann can serve as an example of this attitude. In his book on environmentalism, *Der Tödliche Fortschritt: Von der Zerstörung der Erde und des Menschen im Erbe des Christentums* (The Deadly Progress: On the Earth's and Humanity's Depletion in the Inheritance of Christianity) (Freiburg: Herder, 1991), p. 94, he writes about "Jewish-Christian crossing of feelings." Advocating the original connectedness of humanity and nature, Drewermann proposes the old Egyptian and related worldviews (exactly the ones that were condemned by the Hebrew prophets). In doing so, he blames the (Christian) Jewish tradition for (Christian) bourgeois society's guilt. This is the beginning of the old anti-Semitic song all over again.

9. Of course Germanic, Celtic, and other cultures influenced Christianity as well.

10. I do not think it is possible to link prophetic and biospheric 'faith,' as Daly and Cobb propose. See Herman E. Daly and John B. Cobb Jr., *For the Common Good: Redirecting the Economy Toward Community, the Environment, and a Sustainable Future* (1989; Boston: Beacon Press, 1994), pp. 389–93.

11. Many aspects of this issue are collected by Howard Eilberg-Schwartz, *God's Phallus, and Other Problems for Men and Monotheism* (Boston: Beacon Press, 1994).
12. For example, in the story of Jacob's flight, a chosen person's unique experience is directly linked to 'erected stones' (Genesis 28:10–22).
13. 'Partisan of the poor' is originally Karl Barth's expression.
14. See Isaiah 45:1–8.
15. See Ulrich Duchrow, *Alternatives to Global Capitalism, Drawn from Biblical History, Designed for Political Action* (Utrecht: International Books, 1995), especially pp. 127–229.
16. Matthew 6:33.
17. Genesis 1:26–28.
18. See Ruether, *Gaia and God*, pp. 65f. She mentions the Noachian covenant as a moment when humanity is given more power over nature than in Genesis 1. She links this to the consciousness of judgment. Here again, nature is seen as part of God's covenant with humanity.
19. See Martin P. Nilsson, *Geschichte der Griechischen Religion*, Vol. 2, *Die Hellenistische und Römische Zeit* (München: Verlag C. H. Beck, 1974), pp. 603–4, on Hermetic's and Zoroaster's dualism.
20. G. E. M. de Ste. Croix, *The Class Struggle in the Ancient Greek World: From the Ancient Archaic Age to the Arab Conquests* (Ithaca, NY: Cornell University Press, 1981), pp. 69–80.
21. See W. K. Guthrie, *A History of Greek Philosophy*, Vol. 5, *The Later Plato and the Academy* (Cambridge and London: Cambridge University Press, 1978). See also Martin Heidegger, *Nietzsche*, Vol. 1, *The Will to Power as Art*, and Vol. 2, *The Eternal Recurrence of the Same* (San Francisco: HarperCollins, 1991; German, 1961). In volume 1, chapter 22, "Plato's *Republic*: The Distance of Art (*Mimesis*) from Truth (*Idea*)," Heidegger writes: "what we call 'nature,' the countryside, nature out-of-doors, is only a specially delineated sector of nature or *physis* in the essential sense: that which of itself unfolds itself in presencing. *Physis* is the primordial Greek grounding word for Being itself . . . what is essential in pure Being, as present of itself, in other words, what emerges by itself, stands in opposition to what is produced only by something else" (p. 181).
22. See Hans-Georg Gadamer, "Mythos und Wissenschaft," in *Christlicher Glaube in Moderner Gesellschaft*, Vol. 2. (Freiburg: Herder, 1981), pp. 6–42. The second wave of enlightenment occurs as late as the eighteenth century, "which reached its climax in the rationalism of the French Revolution's era." Gadamer mentions a third wave "in our century's movement of enlightenment, which, for the time being, reached its height in atheism's religion and its institutional foundation in modern atheistic orders of the state" (pp. 8–9). I consider the twentieth century to be a continuation of eighteenth-century Enlightenment, so I am not too pleased with Gadamer's latter characterization. For this reason, I am advocating a third wave of enlightenment, which is critical to "atheism's religion and its institutional foundation in modern atheistic orders of the state" (Gadamer) as well as to the economic-technological system in which we live. See also Gadamer's *Truth and Method* (New York: Continuum, 1994; German, 1960), pp. 200–3, 285–90.
23. See Nilsson, *Geschichte der Griechischen Religion*, Vol. 2.
24. Ibid., pp. 257–68, 395–415. Telling are the words of Stoa's famous representative, the emperor Marcus Aurelius (121–180): "Everything being interwoven, this bond is holy and almost nothing is strange to something different."

25. See Plato, *The State* (*Politeia*), book 6, chaps. xx and xxi.
26. See chapter 4.
27. In *Theaetetus*, a book on epistemology, Plato speaks of Socrates as a 'midwife.'
28. Among Jewish writers, there is a wide variety of interpretations of creation and messianism. Erich Fromm gives a beautiful, psychological-oriented interpretation in *You Shall Be as Gods: A Radical Interpretation of the Old Testament and Its Tradition* (New York: Fawcett World Library, 1966), pp. 96–105. He states: "The messianic time is the time when man will have been fully born. When man was expelled from Paradise he lost his home; in the messianic time he will be at home again—in the world. The messianic time is not brought about by an act of grace or by an innate drive within man to perfection. It is brought about by the force generated by man's existential dichotomy: being part of nature and yet transcending nature; being animal and yet transcending animal nature. This dichotomy creates conflict and suffering, and man is driven to find ever new solutions to this conflict, until he has solved it by becoming fully human and achieving at-onement" (p. 98). Notice the last word, 'at-onement.' Emmanuel Levinas, another Jewish thinker who wrote about messianism, worked out this aspect extensively in his philosophy of the other. See his *Totality and Infinity* (Pittsburgh, PA: Duquesne University Press, 1969).
29. See chapter 2.
30. The ancient empires, preceding fourth century B.C. Greek culture, were already large-scale political organizations, such as the Persian empire, which deeply affected Greek society. Thus, Aristotle's description of the Greek polis is at least partly an idealized historical concept. See Duchrow, *Alternatives to Global Capitalism*, and G. E. M. de Ste. Croix, *The Class Struggle*, pp. 118f., 384ff.
31. See Dietrich Bonhoeffer, "Thinking in Terms of Two Spheres," in *Ethics* (London: Fontana Library, 1964), pp. 196–207.
32. For a fresh, positive view on irony, see Richard Rorty, *Contingency, Irony and Solidarity* (Cambrigde: Cambridge University Press, 1989).
33. Primavesi, in *From Apocalypse to Genesis*, pays much attention to biblical combinations of food and sustenance; see pp. 238–43, "An Ecological Reading of Genesis 1–3."
34. This does not exclude the possibility of a now living person considering himself or herself a member of the Judeo-Christian tradition, as I described in chapter 2. There I described being a member of tradition as 'remembering the great words and commandments'—in short, the major decisions that the Moses tradition (Jesus being the second Moses) made concerning the relation among humans and between humans and nature within the semantic field of *dabar YHWH*, the Word of the Lord. The joint Jewish, Christian, and Islamic cultures' inheritance of being members of the Jewish tradition could be important for our individual awareness of being subjects.
35. These qualifications are derived from a Dutch book on church history, H. Berkhof and Otto J. de Jong, *De geschiedenis der kerk* (History of the Church) (Nijkerk: Callenbach, 1973), p. 34. Ruether, in *Gaia and God* (pp. 122–6), dealing with the dualism of good and evil, describes the relation of Gnosticism and Platonism: "In Gnosticism . . . this hierarchical cosmos of Platonism has fallen into more total dualism."
36. See Ulrich Duchrow's basic study *Christenheit und Weltverantwortung: Traditionsgeschichte und systematische Struktur der Zweireichenlehre* (Stuttgart: Klett-Cotta, 1983). See also Gerta Scharffenorth, *Den Glauben ins Leben Ziehen . . .: Studien zu Luthers Theologie* (München: Kaiser Verlag, 1982), pp. 205–313.

37. E. J. Dijksterhuis, *The Mechanization of the World Picture* (London: Oxford University Press, 1969), (originally, in Dutch, *De mechanisering van het wereldbeeld* [1950; Amsterdam: Meulenhoff, 1985], pp. 447–8, 527–39.
38. Duchrow, *Christenheit und Weltverantwortung*, opposes *civitas dei* and *civitas diaboli* (pp. 181ff.).
39. Augustine is especially influenced by the neo-Platonist thinker Plotinus (third century A.D.). See Karl Jaspers, *Plato and Augustine*, trans. Ralph Manheim (San Diego: Harcourt Brace Jovanovich, 1962), pp. 68–77. See also Duchrow, *Christenheit und Weltverantwortung*, pp. 187ff., 200ff.
40. Ruether, *Gaia and God*, p. 74, writes: "Augustine's own vast treatise on world history, *The City of God*, incorporated the apocalyptic dramas of the Hebrew and Christian scriptures and thus assured their continued importance in the medieval church."
41. See Jaspers, *Plato and Augustine*, p. 99: "The history of mankind is the story of Creation and man's original estate, of Adam's fall and the original sin that came with it, of the incarnation of God and the redemption of man.... The intervening history is essentially of no importance. All that matters is the salvation of every soul."
42. Duchrow, *Christenheit und Weltverantwortung*, pp. 407–8, equalizes the oppositions in Aquinas's scheme: divine law and human law, supernature (grace) and reason, *lex divina* and *lex humana*, *coram deo* and *coram hominibus*.
43. See Jürgen Moltmann, *The Trinity and the Kingdom*, trans. Margareth Kohl (San Francisco: HarperCollins, 1991), especially Chapter 6, "The Trinitarian Doctrine of the Kingdom."
44. See Paul E. Sigmund's introduction in *St. Thomas Aquinas on Politics and Ethics* (New York: W. W. Norton, 1988), pp. xiii–xxvii. See also Primavesi, *From Apocalypse to Genesis*, pp. 100–10.
45. This makes Aquinas an obvious opponent of Joachim of Fiore (see chapter 1).
46. Duchrow, *Christenheit und Weltverantwortung*, pp. 441ff., 479ff., discusses Luther's swallowing the eschatological Augustinian tradition as well as the eschatological tension between God's field of power and the devil's one.
47. Max Weber, *The Protestant Ethic and the Spirit of Capitalism* (1905; New York: Charles Scribner's Sons, 1958).
48. Dietrich Bonhoeffer concisely formulated these problems in 1944 in *Letters and Papers from Prison* (New York: Macmillan, 1972), especially the letter dated July 16.
49. See Ian Barbour, *Ethics in an Age of Technology* (San Francisco: HarperCollins, 1993), pp. 3–56.
50. See chapter 5.
51. Gianni Vattimo, *The End of Modernity: Nihilism and Hermeneutics in Postmodern Culture* (Baltimore: John Hopkins University Press, 1988), pp. 21–2. In Vattimo's opinion, Nietzsche and Heidegger approach nihilism in the same way: "For Nietzsche the entire process of nihilism can be summarized by the death of God, or by the 'devaluation of the highest values.' For Heidegger, Being is annihilated insofar as it is transformed completely into value" (p. 20).
52. Levinas, *Totality and Infinity*.
53. This is discussed by John K. Galbraith, éminence grise of Left-liberal economists, in his book *The Culture of Contentment* (London: Penguin Books, 1993).
54. This is a core theme in Levinas's ethical thought. See his *Totality and Infinity*.

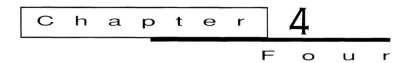

POPULATION, LIFE, AND DEATH

THE HUMAN CONDITION IS a delicate balance of life and death. Human sexual relations and the use of contraceptives certainly affect the creation of new life, but when we discuss a population at large, there are other factors as well. In deep discussion, we must acknowledge a population's position in the life-death balance. The current 'population explosion' is part of the twentieth century's problem with the life-death balance. Deep economy is central to this problematic issue. I begin by focusing on the Latin American liberation theology discussion. This discussion includes the theme of life and death, often in a deep economic context. This is important, because the continuous working of economy is implicitly connected to current population problems. To understand our current situation in this area, we must again trace the history and development of modern traditions. Costa Rican economist and theologian Franz Hinkelammert describes Christianity's dilemma with its Hebrew and Greek roots.[1] The sacrificing of children is present in both of these traditions. In order to understand today's population problems, we must examine historical and current conceptions of 'child sacrifice' and relate them to the individual and collective experience of life and death.

SACRIFICING CHILDREN

Two conflicting traditions compete for priority: the Hebrew 'faith of Abraham' tradition, and the Greek 'Occidental Iphigenia' tradition. The Hebrew tradition features a refusal to sacrifice the child. The Greek tradition features a willingness to do so. The solution to the conflict between these two traditions became the very heart of Christianity: The Heavenly Father sacrifices Christ, the Son. Christians believe that this sacrifice ended all other sacrifices.[2] But has this stated religious truth been a principle in the daily practice of Christian society?[3]

Tracing our tradition's roots helps us understand the deep character of our culture. Ancient child sacrifice myths provide us with insight regarding the moral and religious background of the global treatment of children in the twentieth century. The increasing evidence of widespread child abuse confirms an alarming characteristic of modern society. The way in which children are treated is a major component in the current population problem. The following conviction has considerable validity in the realm of social experience: "Human society is built on a foundation of myths, which are the origins of social consciousness, and which penetrate all social relations—especially relations of power and dependency."[4]

The widespread acceptance of child sacrifice is evident in high birth rates resulting from high mortality rates. In the context of deep economy, the child is being sacrificed for the well-being of the wealthy world powers. The ideology of the economically powerful Western world is responsible for the current situation and practices. Many environmental ethicists in the Malthusian tradition blame the Third World for the situation on the grounds of irresponsible behavior. It is important to realize that this is not the case. The Western world wields the power and ideology, and the blame belongs therein. The Western readiness to sacrifice in general was clearly demonstrated in the so-called lifeboat ethics (which espoused 'spaceship Earth' metaphors), which was based on modern 'natural selection.' Growth is considered the crucial ecological problem; thus the weak should be sacrificed to save the strong. Lifeboat ethics considers it a moral decision to stop helping poor people and societies. The moral argument is based on the presupposition that too many inhabitants would overburden the earth's carrying capacity.[5]

Western culture incorporates two opposing traditions of child sacrifice: Abraham and Oedipus/Iphigenia. This combination casts a revealing light on the moral aspects of the current 'Malthusian dilemma.' Both traditions originated in primeval stories about violence. Violence is at the origin of almost every culture. In fact, culture is a solution to the problem of coping with the life-threatening phenomenon of violence.[6] The moral standards that a society employs to deal with the problems of power and violence are evident in the culture's narratives on the relationship between father and son.

Abraham, the Hebrew patriarch, does not kill his son, even though his God has told him to do so (Genesis 22:1–19). The original interpretation of the narrative focuses on Abraham's obedience; Abraham expresses a willingness to sacrifice his son, and this is what his God requires. However, another interpretation elicits the very opposite requirement from Abraham's God. Here, Abraham's decision arises from his faith in YHWH. YHWH commands Abraham to abandon the tradition of sacrificing children, which was a widespread practice in ancient societies. Thus, Abraham disobeys the common divine 'natural law' and follows the new, different God. This new faith makes Abraham a stranger in his own land. As a stranger and a sojourner, Abraham obeys separate laws, not the established laws of the region. It is easy to see the Abraham story as an introduction to the heart of Israel's faith and culture: living according to the Torah in a promised land.

'Abraham's faith' became a traditional password for a posterity that refused to accept the killing of children and other vulnerable beings. In the New Testament, Paul's focus on Christ's bodily resurrection is understandable in terms of Abraham's faith. As a Jew, Paul lived under the oppression of the Roman Empire. Roman officials killed Jews, such as Jesus, and other vulnerable people after fake trials. The killing was an effort to erase all resistance to the Roman system, and it was executed in the name of imperial divine order. Paul's Easter (Resurrection) message criticizes all killing that is supposed to save life, and it assures that the killing will end.[7] Paul's message on the resurrection of Christ repeats the Hebrew meaning of Easter, which is the exodus from Egypt. The exodus meant an end to the killing of Hebrew children. In the Exodus narrative, Abraham's faith is experienced on a large scale.

Christianity incorporates both the Hebrew and the Greek tradition. We can already see that the New Testament includes Abraham's faith, which denounces the killing of children. Stories in the Greek tradition, however, contradict the message of the Abraham, exodus, and bodily resurrection stories. Instead of denouncing the slaughtering of the weak and advocating the abolishment of laws that uphold such practices, this Greek tradition condones the sacrificing of children and precious objects to propitiate the gods. Although Helen is usually credited for beginning the Trojan War, we can actually trace the war's origins back to Agamemnon's sacrificing of Iphigenia. To propitiate the god of the wind, the Greek king sacrificed his daughter. This sacrifice enabled the Greeks to reach Troy, which ultimately, as Homer describes in *The Iliad*, led to Greek victory.

Aeschylus and Euripides both dramatize this story of sacrifice and victory, stressing the tragic element's free acceptance.[8] In Aeschylus, the daughter protests her father's offering, but in Euripides, she exclaims: "Kill me, destroy Troy . . . servants are they, we are free."[9] Thus the sacrificing of a child is fully accepted; even the child agrees to the offering. Her victimization symbolizes Greek freedom, as opposed to the Trojans' slavery. In accordance with this train of thought, freedom legitimizes the sacrificing of a child. This symbolic narrative is also a comment on Greek superiority. The Greeks accept fatal destiny, which consequently frees them and allows them to exhibit *ataraxia* (imperturbability), impeccability, and moral superiority. Greek *ataraxia* eventually becomes the Stoic ideal.

In Greek drama, characters frequently wrestle with the concept of destiny. The Oedipus story is a famous example. Freud reads the story as an archetype for the son's desire to kill the father. In the Greek myth, however, the archetype is actually the other way around. The father learns from the Delphian oracle that his son, Oedipus, will kill him. By killing Oedipus first, the father tries to escape his own destiny. The violence initiated by the father leads to the destruction of the entire family. In contrast to Abraham, the father tries to preserve his freedom and save his own life by killing his son.

The tradition of Greek tragedies advocates living in accordance with one's destiny, which entails an acceptance of destiny's potential dark side. Humans cannot escape the caprice of nature. Thus the Greeks conceived of natural forces as gods whose will determined human fate. From a psychological perspective, we can state that Oedipus's father was trying to ensure his own safety by eliminating risks. The Greek story demonstrates that destiny is determined. In the Abraham story, however, the future is undetermined. In this tradition, halting the killing offers humanity the chance of a good future.

According to Hinkelammert, Western culture has chosen the Oedipus-Iphigenia tradition; it accepts the practices of killing and sacrifice. The Abraham tradition proved to be too threatening for Western culture. In terms of deep economy, this means that in order for Western culture to protect its wealthy, life-sustaining conditions, it must disregard unfruitful life. In religious terms, this means that if the economic father requires sacrifices, we had better hand them over. Consequently, if the economic prediction is that there is not enough life sustenance for a certain population, we should not intervene in the population's demise.

All this is integral to the long, historical tradition of child care and child abuse within cultures. The choice is always between Abraham's faith and the Occidental Iphigenia and Oedipus tradition. The typical environmental ethicist's approach (the 'Malthusian dilemma') does not effectively reach the heart of the situation. This approach tallies up the numbers, be they birth rates or mortality rates, and exclaims that something should be done. In order to do something, however, we need to know how the situation was created and how it is being sustained in the world today. The situation is less about numbers than it is about the underlying beliefs and ideologies that have propelled the numbers to their current proportions.

Choosing the Occidental Iphigenia and Oedipus tradition means accepting the present global inequalities as fate. To take this path means making an effort to reduce birth rates and accepting the process of 'natural selection.' Human rights violations and large-scale economic violence are condoned as unavoidable, albeit unfortunate, results of living in a 'survival of the fittest' type of world.[10]

Choosing the alternative tradition, Abraham's faith, offers a different set of possibilities. Taking this path would mean rejecting the free-market economy's 'fatal attraction.' This fatal attraction is defined as the sacrificing of part of the world's population and nature's richness in the interest of the remaining part's wealth and well-being. In practice, this postpones the richer person's death by hastening the death of the other. Abraham's faith means not letting the children die off (passive) but feeding and nurturing them instead (active). On the scale of global relations, this path requires a change in the distribution of wealth and inclusion in global economic activities.

These moral considerations focus on personal choice. In history, there are communities of people who have made choices on the basis of age-old discussions. These people formulated their identity as subjects in their tradition, and this included being responsible for their choices. By virtue of the act of choosing, these communities denied the conceptual path of fate and destiny. This attitude, however, does not provide an easy solution to the modern crisis. It forces people to examine fate closely (*fatum* is defined as the course of things one cannot influence). Abraham faces fate, whereas Agamemnon accepts it. Facing fate means examining the established condition and deciding whether it is in fact right. Facing fate is the beginning of liberty, and the evolution of the Abraham story throughout the Old and New Testaments demonstrates this. Once fate is faced, the struggle for substantial changes can begin. This process is the essence of enlightenment. Yet an enlightened or emancipated society can proceed to erect new systems that prove to be as fatal as the previous ones in certain ways. Modern economy is an example of fatal enlightenment. So we find ourselves at a point where the whole process of 'disenchantment' should start anew.[11] When we approach the potentially fatal problem of the population crisis, we need to face fate by examining the current state of human rights. And in the tradition of our forebears, who critically examined 'divine representation' and 'natural law,' we must hold the economic father up to rigorous scrutiny.

POPULATION AND REPRODUCTIVE RIGHTS

There was a perfect balance between slave and master in ancient Greek society, especially in the sphere of *economia*, the work in the house (*oikos*). In politics (the public realm of the polis), the existence of the slave was denied. This was a complicated way of avoiding dichotomy among people in the interest of saving the real community.

Aristotle compares the cooperation between master and slave with the bond between man and woman, the lord (*oikodespotes*) and his wife. Both relationships are hierarchical and are considered natural. In Aristotle's view, children are the natural result of the harmonious male-female relationship. Man and woman each have a proper place and task in the process of begetting and raising children. In the familial structure of the *oikos*, all participants have certain rights and duties, assuming that no one contests the delegated position. This hierarchical and metaphysical structure delegates positions and demands virtuous behavior. The polis acknowledged women and children as humans but limited public administration to free male citizens.

Today we oppose this harmonious small-scale concept of reality and opt for large-scale social and economic processes. These processes cause people to live separated from one another. Anyone who wants to analyze our current population problems must start by examining the division between the sexes. The birth-rate problem and population policy stem from male power. The patriarchy structured the global economy, and it largely determines the way in which women arrange their own lives.

Recently, female authors have advocated an approach to the population problem that differs from the male approach, which is a technical concentration on birth rates and nature's carrying capacity. These women argue that the population issue is a question of begetting and giving birth to children; thus it is a women's concern. This position is forcefully expressed with the term 'reproductive rights.'[12] Betsy Hartmann uses this term to confront official reports and strategies that often display the West's attitude of superiority in relation to Third World countries. This confrontational term alerts the reader to the patriarchy's refusal to deal with the population problem as a human problem. Instead, the patriarchy uses the term 'procreative agents' to describe Third World people.

Els Postel's analysis of the United Nations' reports on population issues charges that only technical questions are asked, and in response, only technical solutions are suggested.[13] Apparently, the United Nations thinks that the population problem is a technical problem, which is typical of the scientific reductionist tradition. Its central question is which effective contraceptive, including sterilization, will people willingly use? And if they refuse contraception, should they be forced? How could such a policy be enforced? History predicts that women would be the first targets in a policy of forced contraception.[14]

This is a large-scale example of the male power structure's oppression of women. We can see how power relations are applied in practice when we localize this oppressive power relationship in a deep economic context. And conversely, deep economics is not only economic in its nature. Its deep character includes male and female interactions, as well as the creation of 'new' and the extinction of 'old' life.

Regarding the population issue, we encounter large-scale demographic theories and scholarship on the smaller scale of interhuman relations. The policies of the United Nations, as well as those of individual nations, often combine these separate spheres. On the basis of large-scale data, measures are applied to small-scale relations. Due to this method of misapplication, practical effectiveness is hardly possible. These separate fields are summarized in terms of the following characteristics:

- Small-scale relations include cultural aspects, gender aspects, male domination, women's reproductive rights, contraceptive feasibility and acceptance, and the personal conception of the 'happy family.'
- Large-scale demographic theory and practical policy include public information and education, legislation, establishing measures for discouraging large families, presenting birth-control programs, and finding the means to execute those programs.

These characteristics operate in an increasingly negative way, because the Western dualistic attitude establishes an unequal opposition: the public sphere (large scale, First World) versus the private sphere (small scale, Third and Second Worlds). Moral pleas for right behavior in the interest of the common good in small communities fail because large-scale injustice is devastating to small-scale relations. Rosemary Radford Ruether comments on this:

> The extreme maldistribution of access to land, jobs, services, and education today means that populations with low consumption and technology will most likely not be living in a healthy simplicity, but in misery and degradation. Thus, the high consumption of the wealthy few and the low consumption of the many are not separate, but interdependent, realities. The same system of power that allows a small percentage of the world's population to monopolize most of its resources also throws the growing masses into conditions of misery and into an environmentally destructive relation to their habitat.[15]

The key word in this quotation is 'access.' Ruether does not ask the deep economic question: Why do some groups have access to all sources of wealth and other groups have no access to it? However, her accusation is correct; this is how small-scale communities experience the global economic system. All efforts to make large-scale decisions work in local communities should be evaluated with this situation in mind. Large-scale injustice is economic in character.

The human part of decision making on all levels should not be neglected. Within the established structures, humans write policies and design social development programs, including birth-control directives. We must ask: Who is making policy for whom? The fact is, most political decisions are made by men, who think in terms of technical measures and solutions. When they apply these to small-scale life environments and communities, female interests are frequently disregarded and violated.[16] The major population issues, such as giving birth to and nurturing children and using measures to stop the 'production of new life,' belong to the female dominion. Thus, the female dominion is where large-scale and small-scale relations should connect. Men's power over women should be confronted openly within the small-scale sphere. Men should respect women's lives, rights, and positions.[17]

Of course, I am not naive enough to think that First World officials will soon begin living among Second and Third World women in the interest of gaining a better understanding of their lives. Yet I do think that the realization that population problems are directly linked to women's rights might weaken the popularity and supposed superiority of technical solutions. A solution to the population problem requires a discussion between the involved parties in which full respect is paid to the validity of female dominion in the situation. This discussion would inevitably reveal the pitiful economic relations between rich and poor countries. My aim is to show the strong link between large-scale policy and small-scale consequences. Yet I must acknowledge that discussing the population problem within the framework of male-female relations and the balance between life and death raises moral issues concerning sexuality, abortion, euthanasia, suicide, war and peace, poverty and prosperity, and so on. For the moment, however, I am putting all these issues aside and concentrating solely on the parallels between the creation of money and the creation of new life. Ultimately, this approach will enable a clearer perspective on these ethical problems. When the separate realms of individual interest and the common good have been brought together, the subject's morality will have changed too. This is due to the fact that there will have been a change in the framework for ethical reasoning.[18]

In the most basic terms, we can say that humans are profoundly relational beings, which consequently generates new life and means of life. This is ancient wisdom. Aristotle literally compared the union (coitus) between man and woman with the working relationship between master and slave.[19] The first form of 'cooperation,' man and woman, assures life's genesis (the continuation of the generations); the other, master and slave, assures life's preservation (sustenance of existing life). Man and woman propagate their kind; master and slave propagate life. What sort of relational behavior do we really have in mind when we discuss population problems? Cooperation? Exploitation? Equality or power basis? Small or large scale? The creation of life is inextricably bound to the means of sustaining life, so we should include both creation and maintenance in any population problem discussion. In addition, we should notice that both require a level of cooperation. In Aristotle's description of societal relations, he assumes that everyone has access to resources for life maintenance. The fundamental difference between Aristotle's 'creation of life sustenance' society and modern society is that the creation of money obstructs numerous people's access to life sustenance. This modern development has provoked a profound change in life and death patterns. Once the balance between life and death was disturbed, population growth got totally out of control.

Everyone agrees that something has to be done. Adam Smith's approach of adapting morals to the situation is clearly not a solution. Smith intended the right moral attitude, such as sympathy, to enable successful participation in commercial society. In application, this approach discourages the poor from creating new life because their lives are already so miserable; this implies that getting rid of children will improve life conditions. Interestingly, experience has proved that when there are favorable socioeconomic conditions, people are more receptive to family planning and decreasing birth rates.[20] So in this instance, large-scale decisions positively influence small-scale life.

THE CHANGE IN LIFE AND DEATH PATTERNS

Economy is a matter of life and death. The delicate nature of the balance between life and death has been illustrated by the numerous cultures that have had extreme difficulties in managing it. The tragedy of contemporary solutions is that the precarious character of the balance between life and death has been forgotten. This is true in both the environmental sphere and the sphere of human justice. It holds true in wealthy and in poor parts of the world. The disturbance of the balance is anchored in industrial capitalism. Environmental problems are a rather late effect but are nonetheless a direct result of industrial capitalism. The staggered cause-and-effect relationship has contributed to our general loss of consciousness about the delicate natural balance between life and death.

Over the last centuries, a completely different form of human living on earth has been established, especially in the Western world. Current population problems cannot be compared with problems from earlier times. Through many advancements in food production and distribution and in medicine and hygiene, the maintenance of existing life is now stressed rather than the replacement of life by new life. This may sound distant and pragmatic, but a major development on both the material and the spiritual levels is distinguishable by using these terms.

In Western culture, a lengthy process has led to important choices about how to treat life, mortality, and death. The inevitable makes way for the avoidable; fate fades away. What does this new paradigm mean in the framework of the environmental discussion and the experience of mortality? I am intentionally not speaking of death. Death is merely one aspect of mortality—an important one, surely—but mortality is a more encompassing notion. Mortality comprises both life and death, and new life is also a part of mortality. The idea of being able to control and manipulate mortality is one of Western culture's major conceptions. Risks can be limited, and it is possible to improve the basic human condition. With the systematic study of reality, causes and effects become facts rather than mysteries. Thus we no longer have a place for a natural higher power, a 'provider.'[21]

We now have to organize a society for an increased number of people, people with changed life expectations and different views concerning human rights and standards of living. We can see the transition from a life attitude of passing through and replacement to an attitude of holding on and preservation. This change is undoubtedly related to the position of the individual, autonomous subject. Individuals consider themselves unique and are hardly able to imagine their own deaths. Death is the one certainty in life, yet this knowledge does not have to be an integrated part of the way of life. The Western world has created a problem with the maintenance and the preservation of life—namely, the individual's own. This ideology fits perfectly in deep economic practice; individuals and groups pursue their goals at all costs, even at the expense of other life, as long as the benefits seem to outweigh the costs. From the Western viewpoint, the overpopulation problem is not that taxing: In order to give every individual a fair chance and share, we have to reduce the number of births.

Yet a major dilemma arises when we acknowledge that large parts of the world neither practice birth control nor plan families. What about the countries where typical Western goods are rarely, if ever, found? In the general human and environmental interest, drastic birth-control measures need to be prioritized here too. Yet the conditions for this to happen are completely different from the ones in the rich countries. To the above-mentioned arguments of large scale versus small scale and male power versus women's interests and rights, we can add the phenomenon that the traditional attitude of passing through and replacing life still holds strong in many poor countries.[22] In countries where, traditionally and circumstantially, Western individuality does not preside over forms of community and kinship, the creation of new life is a given. The creation of new life concentrates on passing through and replacement. This means that a new life is created, it passes through, and it is replaced by another new life.

There are different reasons for the need to create new life. Of course, all species have a basic drive to carry on the genesis of their kind. We live on through our children, and we stand in a tradition of generations; this is an important and cherished value in every culture, and it has always been connected to a social aspect. Parents raise children, who reciprocate the caretaking when the parents become elderly. In this sense, children are insurance for the parents' old age. At the same time, having many children is insurance against death. In a society where infant mortality is high, having many children is a must; the strongest survive to carry on the culture and care for the elderly. In these societies, children are a type of capital; one needs many to face calamities and recessions.

These social aspects are colored by time and place. In many Third World countries, the scenario depicted above is the everyday reality. The number one cause of high infant mortality is harrowing poverty and drastic climate changes. These causes are direct consequences of global ecological and economic problems. Thus a vicious circle is created. Many children are produced to combat the high mortality rate, but because the large number of children cannot be sustained, the high mortality rate perpetuates itself.

The popular Western reaction to high birth and mortality rates is to quickly halt development aid to poor countries. The West believes that such aid interferes with the natural processes in these countries. This is known as lifeboat ethics, and it too perpetuates the vicious circle of high birth and mortality rates in less fortunate parts of the world. Yet lifeboat ethics also preserves the superior quality of life in the First World.[23] Although these attitudes toward mortality are partially aligned with cultural and religious traditions, the large-scale and crucial point is the current division of money, knowledge, and power. It is this division that puts birth and mortality rates on the map of global concerns.

GAINING BY DESTRUCTION

This chapter began with an examination of violence as a religious phenomenon. In ancient Greece, children were sacrificed to the Gods within the context of war and fate (Oedipus and Iphigenia). The ancient Hebrew culture refused to continue this tradition of sacrifice (Abraham's faith). A hermeneutical connection was established between the religious character of modern times and ancient Greece on the basis of our current system's demand for vast sacrificial qualities of humans and nature. Thus we must compare the creation of life and the creation of money on a deep level. In contrast to the romantic ideal of love and organic cooperation as the genesis of creative acts, we find that destruction plays the dominant role in the modern creation of new life and money.

Freud identified the human as a being dominated by hidden passions and desires. Freud found Eros (sex) and Thanatos (death) to be original passions, or basic instincts. His findings enabled us to explain aggression and the urge to destroy. The well-known phenomenon of sexual aggression informs us about the relation between violence and the conception of new life. Violent occurrences have been documented in various environments and situations: from family life to warfare, from incest and spousal rape to systematic rape during wartime. I am suggesting that an indirect connection, not a direct connection, exists between destructive sexual violence and the conception of new life.

Humans are aware of the temporal nature of their own lives. They know that they are destined to die, but they also understand that reproducing continues human life.[24] Humans must cope with this knowledge of their existence, and they do so in confrontation with other humans. Hence, we can consider the vital driving forces of Eros and Thanatos to be basic human instincts indeed. The inclination to improve, defend, or continue one's own life includes violence toward the other, and this phenomenon is easily extended to the clan, group, or community level.

Feminist thinkers have made important contributions in this ongoing discussion. Their focus has been on male violence and destruction. In our patriarchal culture, violence and destruction have been predominantly male activities. However, it is quite another thing to state that males are ontologically aggressive and violent, whereas females are ontologically nurturing and sustaining. We can identify deep cultural and social characteristics as being the products of patriarchal rule, and we can discuss them as such. But until ontological characteristics are indeed facts, we should be reticent to discuss patterns of behavior as belonging to ontological structures.[25]

Translating this relation of violence, destruction, and creation into the field of deep economy, we can say that violence against nature and human life aids in the creation of money and eventually destroys nature and life. Since the creation of money entails an ongoing process of production and exchange, there is no limit to the depletion and pollution of the natural environment. On the economic front, growth and progress demand the abandonment of outdated utilities and services and the accompanying production of new ones. These innovative processes require new types of energy and raw material. Thus these processes require depletion and destruction. For this reason, the term 'capitalization of nature,' often referred to as 'transubstantiation,' is a synonym for the destruction of nature.

The destruction of life affects natural species and their replacement. Modern economic action replaces natural richness with artificial products in the interest of increasing human comfort and wealth. Yet this human action does not provide nature with the chance to replace the depleted or destroyed elements. This is human exploitation of nature. Deep economic research has revealed that the current Western standard of the maintenance of human life requires the exploitation of nature. Science, technology, and economics all cooperate and support the processes of innovation, progress, and growth. Because all these processes are geared toward the development of new things, all antiquated things, policies, and strategies have to be discarded and replaced.

The imperative to discard and replace is part of a conviction that is in accordance with economic laws. Most people are unaware, often intentionally so, of the economy's catastrophic operating character. Deep psychology (psychoanalysis) makes claims for the overcoming of repressed elements. Why shouldn't we apply this claim to deep economy? Deep economic investigations could bring these repressed economic elements to the surface of societal consciousness. An acute awareness could urge a kind of *metanoia* (conversion), which would spur a corresponding change in political and moral agency.[26]

The balance between life and death is our biggest environmental problem. The earth dies because the conditions of life are systematically undermined. Our reality is ironic, in that life-stimulating action causes death. The natural balance between life and death is disturbed by life-stimulating action. From the human's standpoint, it is a slow process. Small changes and disturbances go unnoticed. Yet these small changes form a lengthy chain on a growing scale of disturbance. This explains why significant environmental problems were not identified until the present age. The progression, however, has been going on for centuries. And when we qualify this progression in terms of the earth's life span, rather than in terms of the human's life span, significant environmental damage has occurred at an alarmingly rapid pace.

There have always been movements and institutions that tried to slow down this destructive progression. Their typical methods include an idealization of the past and a defense of traditional doctrines. The most powerful of these groups has been the Roman Catholic Church. Yet none could foresee the enormity of the devastation; today's scale of destruction could not have been imagined a generation ago. And because of the emphasis on positive progress, attention was directed toward positive results, leaving negative effects on the taboo sidelines. The slow process of change toward a better future did not admit the development of a critical attitude.[27]

We see this again and again in Western history. The quick rise of Nazi Germany did not provoke fundamental critique at the time. Germans hoped for the realization of a Third Reich and the establishment of Germany as Europe's leader.[28] In general, when the principle of hope is incorporated into a doctrine of historical progress and social improvement, it prevents everyone involved from noticing and understanding the policy's negative attributes.[29]

Fighting death and overcoming its threatening character have been long-term hopes in human history. To achieve this victory, the experiences of and thoughts on mortality have changed over the course of Western history. The more aware we became of our weak position in natural reality, the more we considered progress to be a necessity and a blessing. The leading questions became how we could accept mortality as a human condition without going under as individuals and as a species, and how we could live with the knowledge that we are deficient and incomplete.[30] Of course, this was put in a positive and practical way: how to live a happy and joyful life. Both practical and theoretical answers to this question always amplify the relation between humans and their natural surroundings and conditions.

Disregarding the warnings and wisdom of classical mythology, modern humans persist in the search for power over nature and mortality. More than ever before, humans seem unwilling to accept their own nature as mortal and perishable beings, to accept health and illness, vitality and suffering as parts of the life process.[31] The guiding principle of 'live and let live' has been banished to the annals of history. 'Live at the expense of' is today's working rule. This is not just an egocentric attitude of the 'blasé bourgeois'; it has become a structure, and it generates policies that can no longer be designed in ways that deviate from the new working rule.[32]

This goes right to the ambiguous heart of Western culture. The Western attitude and system are focused on 'holding tight' and 'maintenance,' yet the West simultaneously executes wide-scale destruction. It appears that maintenance exists on a conscious and willing level, and destruction exists on an unconscious, 'admitting' level. The conscious level serves the rich and powerful individual's own well-being; the unconscious level services the other's destruction.

In premodern society, the phenomena of death and dying were integrated parts of society. The premodern human knew: You die today, I die tomorrow. This awareness did not alleviate mortal fear, but it connected death with the process of 'handing through' and 'replacement.' Something is handed through to a person (the literal meaning of 'tradition'), and that person hands something to others. Thus, I replaced and I shall be replaced. I will continue to exist in my deed, which remains memorialized in the community, and in my descendants, who carry on traditions.[33] The decrease of this attitude and the drive to progress are dialectically connected. By 'dialectically,' I mean that these processes mutually stimulate and preserve one another. This implies a change in the position and self-image of the Western human. In contrast to metaphysical medieval sociology, the modern human is not just a part of the native community, which explicitly connects the person to specified social layers. Theoretically, emancipation from determined social structures is available to everyone.

All humans share certain common experiences, and death is one of them. Yet there has been a dramatic shift in modernity's experience of death. Individual death has been separated from the general mortality of nature as a whole. As the notion of fate-companionship decreases, mortality is experienced emotionally as a private matter. Modern humans often refuse to cope with the awareness of death—hence the suppression of death in daily life and consciousness.

This coping mechanism of suppression is a fatal psychological and social development. It destroys the fundamental feeling of connection with others and with nature. The consequence is that connection and aggression, love and hate become equal. Social individualization stimulates the perception of the other (humans and nature) as a threat to our own well-being. The dangers presented by the other are kept out in two ways: seclusion from outer influences and use of the other for personal advantage. The latter necessitates that the individual connect with reality in a new way. Our social structures enable this, because advantageous use of the other is the basis of our communal system.

Individuals must prudently find their places in the system and function as well as possible. This functioning is, according to the logic of the system, focused on the interests of the individual and of the group that is represented in the system. According to this logic (the *logos* of the system), the whole will eventually profit from the particular interest, as stated by Adam Smith in *The Wealth of Nations*. Restrictions are set only when a certain behavior deviates from the common drive to increase the wealth and sustain the system. The fundamental change in the relation between the individual and the group confirms my earlier opposition to the concept of community as an ontological structure, which implies that the individual needs the community. Admittedly, the premodern individual could not survive without the community. In modern times, however, the group needs individuals: The group doesn't really exist without the individuals who have consciously chosen to belong to the group, thus forming a community. In the sphere of activist, political, economic, and religious activities, individuals decide which causes and actions they will support or oppose.

CREATION OF MONEY

In order to understand the nature of money in modern society, we must understand its three simultaneous functions: (1) the means of trade, (2) the measurer of value, and (3) the means of capital accumulation, which is economic growth. Thus it is obvious that the need to create money is an absolute in our present type of society.

Money's function as the means of capital accumulation explains the process of money's creation. Money is needed for the creation of money. Because money's creation is a continuous process, a lot of it is needed on both ends. This is called the formation of capital.[34] Certain elements play together, for instance, competition, innovation, and concentration. These elements support one another as system variables. They are the conditions: The creation of money depends on the subtle equilibrium of these factors. Money can 'grow' in certain places, and hence new money is created. This fertility of money is comparable to a crop's soil; conditions for growth must be nurtured and secured. One of the conditions for the creation of new money is the growth of existing money; thus the growth of money is a continuous, never-ending process. For the growth and creation of money, one must have the following available at the same time and in the same place: capital, technology, raw material, infrastructure provisions, labor force, management, political organization, social organization, and markets within (financial) reach.

This undoubtedly incomplete list of money-creating conditions illustrates the vulnerability of our massive economic system. Money is created only when capital grows. This means that there must be endless innovation in communications and in production and trade processes. Only strong companies in rich countries are able to keep up; the others go under. Common efforts are needed to protect these strong companies from suffering a loss of one of the conditions requisite for their operation. Any measures with even the remote potential to weaken the economy are avoided, even if the cost is environmental harm or human rights violations. Avoidance is an imperative for the economic system's well-being. In a competitive society, the selling price must be as low as possible. This is the nature of a free-market economy. The need to keep production costs down logically follows this principle. Anything that adds a cent to production costs is absolutely avoided; this explains the resistance toward adopting protective environmental measures and moderating the use of raw materials and energy. Money cannot be created without this attitude of avoidance. In our present economic system, the notion that strong companies will eventually spend more to protect the environment is an uninformed, naive notion. As the system currently stands, money invested must yield maximum profit. The system is based on money's indissoluble character of yielding the fruit of more money under stringent conditions.

As long as people succeed in coping with the system's vulnerability, money's accumulating and growth functions provide modern society with a strong economic base. Competition is the system's leading principle, and it has both positive and negative attributes. Its positive attribute is that a successful position in the competition brings power and wealth. Its negative attribute is that this strong position can be weakened suddenly by numerous factors. These aspects are part of competition's character.

Economists often use the body, which experiences both health and sickness, as a metaphor for the economy.[35] They use this metaphor to describe all types of societies and communities. In this conception, every member of the body has a function that contributes to the society's health. Here the metaphor is applicable to the ecological system as well. Sound cooperation among the members keeps the organism alive. All threatening elements must be removed from the organism immediately.[36] The importance of the individual body part is its contribution to the system's well-being. Thus money as a value measurer is used to gauge a person's worth. This is a purely instrumental evaluation of personal conduct. In spite of continuing efforts of business ethicists, trade unions, human rights movements, churches, and sometimes governments to reroute the focus toward human qualities, the system relentlessly focuses on quantity. Ultimately, the human's well-being is inferior to the system's well-being. It follows that the persons or countries that pose any type of risk to the money-creating processes are removed by exclusion from the economic system. This exclusion is based on the assumption that money cannot be created without the presence of capital and positive economic perspectives.

Underlying all this is the belief that the process of production, trade, and consumption must be controllable. Parties involved in the process must agree on the same goal: a good, cheap, and profitable product. This cooperation demands an equal focus and an unequal division of money and power. One of the system's premises is that those who cannot participate are subsequently and systematically excluded from the economic system.

The system's blood, which is money (according to Thomas Hobbes), does not circulate in underdeveloped nations. These countries are outside of the economic system's body. This is because they lack the capital and the favorable conditions to sustain the 'magical' act of creating money. Money circulates in the developed countries because of their purchasing power; here, production and consumption are high priorities. Yet the raw materials and energy needed for this production are taken from the underdeveloped nations, from the places where the blood does not circulate. Without technological and industrial development, natural resources are not of much value in underdeveloped countries. Thus countries that meet the technical, organizational, and infrastructural requirements can extract natural resources from underdeveloped countries and make huge profits, which generate more money in the extracting countries. The financially empowered countries lend money to underdeveloped countries, but since the underdeveloped meagerly produce money and are structurally handicapped, they are not able to earn any interest. The result is increasing financial problems in underdeveloped countries.

This process incapacitates underdeveloped countries from procuring the necessities of life. The affluent conditions that create and justify the 'hold tight and maintain' attitude toward population are nonexistent, and subsequently, so is the attitude in underdeveloped countries. Necessity compels the maintenance of the traditional attitude of 'handing through and replacement.' Thus the conditional aspects for the creation of money largely contribute to the population problem. Money's first function, the means of exchange, is limited to participants in the wealth system. Within the circle of circulation, the system needs many participants with purchasing power. Outside this circle, however, human beings are unwelcome. Yet it is outside the circle that new life is needed in order to replace the lives that are lost. In order to survive at all, the underdeveloped countries must replace the lives that the affluent countries have sacrificed for their economic health and well-being.

From a moral standpoint, the system's exclusion and sacrifice of Third World inhabitants are cruel and inhumane. But from a technical standpoint, that is the logical consequence of the system. If we take this argument to its next logical step, we find that the problem of feeding all the people in the

world is principally a Third World problem. As the already existing population problem increases, a question arises: How will the people who are excluded from the world's wealth be fed and sheltered?

We should not ask how the earth can feed all its inhabitants, but rather
how the system can be restructured for a more just distribution of the earth's
richness. Now that we are asking the right question, we can begin examining
the earth's capacity, which is directly related to the dominant economic system.
Once we see this, we are in the sphere of deep economic questioning.

CREATION OF NEW LIFE

The population issue has been approached from many different angles. In
her book on reproductive rights, Betsy Hartmann accuses Western population
policy of demonstrating contempt for women, particularly poor women.[37]
Hartmann's diagnosis is right, but in order to treat the problem, we must
recognize that this contemptuous practice feeds into a large-scale division
among people. This division is created by the economic system's structural
exclusions. So we must consider how the creation of new life is analogous
to the creation of money. We should start by asking why there are plenty of
children but not enough money. The inherent ideology underlying the question requires attention. In order to answer the question, we must identify the
deep economic reasons for, as Marx put it, capital's 'canine hunger,'[38] as
well as our own reasons for wanting to stop population growth.

The birth of a child and the birth of money are both vulnerable processes.
In each, conception is shrouded with mystery and is laden with idealistic
and doctrinal reasoning. The rigorous morality that accompanies the conception
of a child is matched by Adam Smith's free-market economics dogma. This
type of attention and regulation are usually intended to protect vulnerable
areas of life. Yet no measure of protection can completely safeguard the
organic and economic sources of life. For this reason, people are willing to
fight for the sources' conservation, and thus they strive for protective power
and power structures. This defensive stance leads to the envisioning of potential
enemies. Under certain conditions, a paranoid crusade mentality spreads when
a country feels that its culture, system, or origins are being threatened by an
enemy, be the enemy or instability real or imagined. The Cold War mentality
is a historical example of political hysteria in the economic sphere.

Many cultures display a deep-seated dishonesty in the areas of sexuality
and birth. These cultures absolutely resist social reality in these areas. In
the United States' conservative circles, family values are regarded as a God-
given mandate. Abortion is absolutely forbidden, and they fight to establish
strong laws against it and severe penalties for it. This behavior makes it
easy to deny the actual situations in which conception, pregnancy, and birth

take place. The pro-life movement exemplifies and typifies this attitude of righteousness and denial. According to this movement's dogma, God gives humans life, and humans have no right to stop it; humans are required to honor life as God's gift. The Roman Catholic Church acts as a champion gladiator in this area by encouraging birth with the prohibition of contraceptive methods. The church forbids sex before marriage and the use of contraception within marriage. The supporting dogma is that marriage is a natural structure, and manipulation of conception and birth is forbidden on the basis that God's work is revealed in nature. Thus, reverence for the prescribed moral standards' holy character provokes the systematic denial and condemnation of the way in which sex, conception, and birth actually occur in social reality. There is a tricky type of dishonesty in this obsession and denial behavior pattern. The obsession with sex and pregnancy is overwhelming, and at the same time, the disinterest in other problems is absolute. Sex, pregnancy, and birth are the mysterious, hidden origins of life.[39] The need to protect this vulnerable and essentially unknown realm makes conception issues instrumental in the struggle for moral and religious power and control. The church regards the creation of life as a mystery, but it then proceeds to dictate the absolute truth regarding these God-given burdens as duties. The irony, of course, is that the discernible social reality is denied and the indiscernible areas are regulated with enforced truths.

The conservative view is that sex and birth are connected to the mysteries of life and death, and humans should not interfere in these matters. Humans should do what they are told without question or deviance: marry, beget children, raise them in the family institution. Conservatives even believe that a virtuous family life wards off wrong economic acting.[40]

A problematic issue is why human autonomy should suddenly stop when we reach sex and conception areas. Are there religious or metaphysical reasons that forbid critical investigation of these 'sacred fields'? Are we really protecting the vulnerable character of sexuality and conception when we leave unexplored its prominent components of power, rape, money, and violence? And if we think it best to neglect these behavioral and circumstantial components, does this mean that we think that the pre-modern, community-oriented morality has such power and influence that basic instincts are rerouted or that laws enforce proper behavior?[41]

The answers to these rhetorical questions lie in the inclination to save community's resources with religious and metaphysical claims. Historically, there is a strong doctrinal line of thought that defines creational acts as God's work, and both wealth and children are considered his blessings. Thus, humans mediate God's work through sexuality and labor; they cooperate and hope to reap the benefits, which are God's blessings. Therefore, these areas of human existence were mandated as offices, installed by God the creator.[42]

In the past, people paid a great price by handing the areas of sexuality and birth over to religious, moral, and scientific officials. Doing so took these areas out of the hands of the people and set a cap on premodern shapes of human autonomy, rights, and freedom. Recently, feminist research has revealed much in this area, especially concerning men's success in taking health care out of women's dominion. In premodern society, health care was a women's field, but with the advent of science, men quite forcefully usurped the field and denied women access to its administration.[43] This usurpation and contempt for women are discernible in the contemporary neglect of female reproductive rights. Institutions involved in decreasing population growth routinely approach the problem as if there were no actual people involved, let alone people with rights.[44]

Modern society has routinely used both God and nature as ideological concepts, and it has employed them against people's real enlightenment and the emergence of human rights. By honoring God or nature (the Roman Catholic Church incorporates both) as the absolute giver and protector of new life, human involvement and responsibility are denied. Thus in the areas of their own sexuality and biological involvement in creational acts, humans are denied rights. Only persons who are ordained representatives of the metaphysical realm, persons who have received training and are considered experts (ironically, these professionals usually take a vow of celibacy and thus are supposed to have neither sex nor children), are entitled to set the moral standards and establish the rules for sexuality and reproductive behavior.

Opposite to the Vatican's and other conservatives' exasperating moral and normative public challenges, I put forward the following normative statements as an ethicist who is impressed by the parallels between the creation of new life and the creation of money. It is time for intellectual discourse in these areas to advocate the people's regaining of knowledge and power over their own lives and bodies. People must be trained in food and health practices, and they must learn to make their own decisions and trust their own judgment. The liberation of the individual must extend to the liberation of groups from oppressive structures. Women should be freed from male domination. The Third World should be released from Western tutelary bondage. This means renouncing patriarchal ideology and behavior. The ignorant late-colonial attitude that Western men are spiritually, morally, and intellectually superior to all others must be relinquished.

I am asking the free subject to take responsibility for his or her own fate, and this means taking one's own fate into one's own hands.[45] This is not a liberal reading of Adam Smith's 'well-understood self-interest.' I propose that the subject is the moral agent directed toward the other. Creation of new life is a mystery indeed. But it is primarily a human mystery, not a biological one. The mystery character of creation does not entitle alien 'experts' to dominate this area. We need to rid ourselves of this archaic inheritance from predemocratic times. The time has come to focus on the people who are actually involved in the mystery. These people are the first to experience the coming of an unknown, yet known, human life—a human life that is both part of themselves and another self in its own right. Separation and connectedness are inherent in the birth process. When we dismantle the ideology and institutions that bar human involvement in the creation process, the people who are truly involved will discover the mystery character of birth for themselves. When this happens, we will discover that the creative power at work is not an invisible hand or some other indistinguishable metaphor for a faraway, divine act. We will discover that the great mystery is the other, the vulnerable person who is suddenly there. The birth of a child is a transcendental experience and occurrence, because the parent's egocentrism is rerouted toward the other's dependency, the other's need for attention, the other as a separate being connected to oneself.[46]

I am not making any claims that this development is a law or a self-evident truth. I am claiming that we need 'moral democratization.' The people who give birth to new life should be approached as moral subjects who know what to do. Social and psychological circumstances will not always be ideal; often they will be outright poor, and problems will always arise. Thus an effective protection of the whole birth-giving and birth-planning process is impossible. Yet there is work that can be done in the preventive sphere. People are social beings, and they must learn how to behave, how to build relations and possibilities both materially and spiritually. People have to be receptive, and they must have mature personalities if they are to care for another person.

All this presupposes adequate material and mental circumstances. Population problems should never be discussed without including these fundamental conditions. Familiarizing the population with sexuality and propagation requires a strong base for responsible behavior. Even with responsible behavior, however, at least three problems have to be faced. First, many communities and countries do not offer members the freedom to talk openly about sexual issues, nor do they allow people to make decisions about the use of contraceptives. Second, there are many places where contraceptives are unavailable or unaffordable. These circumstances are common in underdeveloped countries. Finally, sexuality is often dominated by male power. Men do not want to frustrate their sex drive with contraceptives. From his standpoint, she becomes pregnant, so it is her problem. Women's subservient sexual position makes it difficult for them to solve the problem themselves. Abortion then becomes the only solution. Quite often, this solution is illegal. We are now in a vicious circle.

These practical points articulate the population's vulnerable character. This vulnerability, however, differs from the vulnerable nature of creational acts.

The population's vulnerability can be brought into the open; we can discuss it, and we can change the situation for the better. The population does not have to be so vulnerable. We do have a choice here.

A THIRD WAVE OF ENLIGHTENMENT

Anyone who concentrates on the beginning of creation in a deep way ends up being overwhelmed by visions of eschatology (the end). Similarly, thinking about our civilization's apocalyptic features leads to conceptions of our origins. This thinking process is driven by the urge to find a better end than the one spreading out before us. Is our story's 'happy ending' written for only a few select inhabitants of the earth? Anyone who has followed my argument will conclude that we need to make deep economic changes. We must tap new sources of life that assure a healthier life and death balance. These changes need to occur in daily practice and in human psychology. We must embark on a third wave of enlightenment.[47]

The development of Western Enlightenment began in Greece during the fourth century B.C. It was followed by a second wave in eighteenth-century Europe. We now need a third wave of enlightenment that will free us from the second wave's fatal consequences. Both the first and second waves were primarily Greek in character, although secularization is often claimed to be a consequence of biblical, especially Old Testament, thought.[48] Breaking through the corroded interpretations of prophetic Hebrew narrative has the potential of reawakening Western culture to its own treasures. An examination of the interpretation and practical execution of the stories of creation, Abraham, the exodus, and the Resurrection could be a starting point for unlocking the self-created, self-mutilating bonds of Western civilization.

Some will claim that asking for a third wave of enlightenment is equivalent to asking for the establishment of a new Third Reich. It would entail going back to myths and irrationality, a surrender to the well-remembered tactics of Nazi Germany. Those who think this way have misunderstood me. A flight from modernity is the opposite of what I am proposing for our world. We should never give up the attainments of the long process of Western enlightenment, which literally means lighting the darkness. And it refers to rational and emotional illumination. A third wave of enlightenment should be a continuation of the previous waves. But the third wave should overcome the former's destructive consequences and developments.[49]

The job is extremely difficult, because we are speaking about system components that cannot change without affecting the entire system. But this is what enlightenment does. This is why I argued that the second wave of enlightenment provoked a rupture in the scientific, political, and economic sphere of the then existing system. We can argue whether the development began with theoretical concepts or with action and practical models, but the relatedness between theories and practices is undeniable.[50]

A 'seduction of the spirit' is occurring in contemporary Western culture.[51] People are searching for new spiritual energy and sources of life, but they refuse to attack or even affect the system in the process. The system's deep economic roots are not being brought into the light. The New Age movement is a typical example. Although this is a broad movement with many different facets, there are several common features. In their longing for spiritual enlightenment, New Agers often affiliate themselves with deep ecology. In their endeavor to cover the cosmic reality, they tend to neglect material relations, especially deep economic elements. There is also a renewed interest in Eastern worldviews and lifestyles; other non-Christian sources of Western culture and elements of Christian Gnosticism have become popular once again.[52] These movements and interests reveal a profound desire to find a spiritual source that gives life meaning and coherence.

My plea for a third wave of enlightenment is fundamentally different from these popular spiritual and often dualistic quests for life sources and a sense of meaning. I respect these movements' intentions, and I understand the importance of spirituality and meaning in personal life, but no one is considering the actual sources and processes of life itself. We are living in such a place and time that we cannot afford to overlook these things. These sources are decisive for life and death, and the 'spiritual sources' that everyone is referring to derive their metaphorical power from these concrete sources.

In smaller, premodern communities, springs, land, cattle, housing, and family structures are such sources of life. They are easy to localize, and everyone understands that defending them from threatening elements is a life and death matter. In modern society, life-conditioning sources are found in two creational acts: the creation of new life and the creation of new money. In order to protect these life-source 'acts,' we adhere to a dual policy. That policy is: Everyone is withdrawn from a complete (deep) understanding of these acts, yet everyone must contribute to them in the proper way. In obedience to large-scale policy, people act responsibly in small-scale relations. Yet they do not recognize their obedience to the large-scale system; instead, they interpret their actions as being in accordance with directives 'from above.' Their actual allegiance, however, is revealed when they apply their small-scale behavior to the other in the support of large-scale well-being. Birth control is an excellent example of this. People who need many children because of their (small-scale) situation are told to stop because the ecological and economic systems' carrying capacity (large scale) cannot support them.

In premodern times, *homo religiosus* conceived of a transcendental power that protected life and its sources. It was therefore appropriate for that culture to propitiate this natural or heavenly power. The difference between Abraham and the Greek tragedians lies not in the question of this power's existence but rather in the way they discuss and name its transcendent power. Modern society differs from ancient Greek and Hebrew society in foundational and fundamental ways.

As a result of enlightenment's lengthy development, our sources of life are human actions. Humanity took its destiny in its own hands. The life and death balance depends on the right behavior, and it is therefore vulnerable. When humans usurped their destiny from nature, they created a fatal separation between themselves and nature. The right behavior they developed to protect themselves proved hostile to their environment. In ecological thought, balance is the biospheric life condition; it is of the utmost importance.[53] Food chains are part of the balance: One species serves as another's food. The food chain is a necessary life condition. Death is an integral part of the life system. When one species uses too much or protects itself too well, it disturbs the whole's balance. This is what is happening right now. Humans protect themselves by destroying nature. Humans do this by creating money. This creational act carries specific rules, which are analogous to a transcendental religious power. I have illustrated this power and money's fatal character by referring to mythological and psychological data.

The following aspects must be recognized in advocating a third wave of enlightenment. First, we should avoid idealistic concepts and concepts that are merely materialistic. Working in the eighteenth-century Enlightenment scheme, Karl Marx tried to replace idealistic philosophy with a materialistic approach toward reality.[54] Rather than ushering in a third wave of enlightenment, Marx's approach called for the annihilation (*Verelendung*) of the second wave of enlightenment's attainments in the larger interest of a revolutionary establishment of Western progress and achievement. Obviously, Marx's approach is a dead-end road, but this does not invalidate his analysis of the capitalist system. Marx was a deep economic thinker; his wrestling with Adam Smith proves this beyond doubt. I consider him a modern Jewish prophet, representing the tradition of the faith of Abraham.

Second, we must nurture a spirituality that avoids New Age dualism and deep ecology's one-sidedness. Humans stand in the balance between life and death, and thus they require a vital force in order to survive. Individual involvement with other humans and other species prevents culture from the monomaniacal obsession with the individual's private interest. Unless the individual becomes involved with other humans and species, the materialistic and spiritual self-interest—albeit veiled in lovely rhetoric about wholeness, community, and universal experiences—will continue to pursue its fatal course.[55]

Third, Western culture should be aware of the seduction of 'the image.' Watching television differs from reading a book. Watching television involves looking; reading a book involves thinking. Enlightenment means respect for, attention to, and an interest in thought and critical deliberations. Humans are gifted creatures, because they have the ability to consider their culture from a distance. Every individual is potentially able to do this. It is the condition for being a moral subject. The individual as subject is an important concept that we should not give up in a burst of postmodern fashion.[56] The condition for practicing an enlightened existence is that the human moral subject keeps reading, thinking, speaking, and writing. These activities are the starting points for the organization of enlightenment-supporting groups and communities.[57]

Fourth, a culture should avoid the general moral discussion that is based on idealistic concepts that are derived from the Western tradition's dualistic character. The field of ethics abounds with discussions and deliberations among autonomous individuals. However, this concept should include the participants' positions and their readiness to discuss their own positions in depth.

The involved people's material and spiritual conditions should be regarded as an absolute priority in all areas, and it would be wise to apply this basic principle to the population issue immediately. Enlightenment is considering the whole person, body and soul. Enlightenment applies to individuals, to groups and communities, and to countries' global relations. Enlightenment means cooperation and togetherness in achieving the human and ecological interest of restoring the balance between rich and poor and between life and death.

NOTES

1. Franz Hinkelammert, *Der Glaube Abrahams und der Oedipus des Westens: Opfermythen im Christlichen Abendland* (Abraham's Faith and the Occidental Oedipus: Sacrificial Myths in the Christian Occident) (Münster: Edition Liberación, 1989).
2. See René Girard, *Le bouc émissaire* (The Scapegoat) (Paris: Bernard Grasset, 1982), chap. 9.
3. See Franz Hinkelammert, *Die ideologische Waffen des Todes. Zur Metaphysik der Kapitalismus* (The Ideological Weapon of Death: On Capitalism's Metaphysics) (Münster: Edition Liberación, 1985), pp. 163–91.
4. Hinkelammert, *Der Glaube Abrahams*, p. 18.
5. See Robert C. Paehlke, *Environmentalism and the Future of Progressive Politics* (New Haven and London: Yale University Press, 1989), pp. 41–76.
6. See Girard, *Le bouc émissaire*.
7. See Hinkelammert, *Die ideologische Waffen*, pp. 165–70.
8. Aeschylus, *Oresteia: Agamemnon*; Euripides, *Iphigenia in Aulis*.
9. Hinkelammert, *Der Glaube Abrahams*, pp. 89–91.
10. Hitler's selections were a large-scale caricature, revealing some ultimate truths about modern society.
11. In Max Weber's writings, the term 'disenchantment' (*Entzauberung*) is important. See Eric Voegelin, *The New Science of Politics: An Introduction* (1952; Chicago: University of Chicago Press, 1987), pp. 13–23.

12. See Betsy Hartmann, *Reproductive Rights and Wrongs: The Global Politics of Population Control and Contraceptive Choice* (New York: Harper and Row, 1987); Els Postel, "Gender, Health and Population Policy," *VENA* 3, no. 2 (Nov. 1991): pp. 4–7; J. A. Gupta, "Women and Health: Fertility, Reproduction and Population," *VENA* 3, no. 2 (Nov. 1991), pp. 17–21. (*VENA* is a quarterly publication of the Women and Autonomy Center, Leiden University, P.O. Box 9555, 2300 RB Leiden, The Netherlands.)

13. Postel, "Gender, Health and Population Policy." Postel does discover some improvement, but still very marginal.

14. Hartmann, *Reproductive Rights and Wrongs*, pp. 65–72.

15. Rosemary Radford Ruether, *Gaia and God: An Ecofeminist Theology of Earth Healing* (San Francisco: HarperCollins, 1992), p. 89.

16. Hartmann, *Reproductive Rights and Wrongs*, p. 188, provides us with a striking example: The contraceptive Depo-Provera was suited for men's use but was rejected because of its antilibido effect. It was accepted for women's use, even-though it has the same effect on them, which is apparently unimportant.

17. And men should not speak about 'couple's decisions,' as United Nations reports do, presupposing that couples are a harmonious unit. See Postel, "Gender, Health and Population Policy."

18. See Ian Barbour, *Ethics in an Age of Technology* (San Francisco: HarperCollins, 1993), pp. 110–2, 258–67.

19. Aristotle, *Politica*: "We shall, I think, in this as in the other subjects, get the best view of the matter if we look at the natural growth of things from the beginning. The first point is that those which are incapable of existing without each other must be united as a pair. For example, (a) the union of the male and female is essential for reproduction; and this is not a matter of *choice*, but is due to the *natural* urge, to propagate one's kind. Equally essentially is (b) the combination of the natural ruler and ruled, for the purpose of preservation. For the element that can use its intelligence to look ahead is by nature ruler and by nature master, while that which has the bodily strength to do the actual work is by nature a slave, one of those who are ruled. Thus there is a common interest uniting master and slave" (1252. a. 24–34).

20. See Hartmann, *Reproductive Rights and Wrongs*, pp. 271–85; especially pp. 278–9, or the Korean 'miracle.'

21. Deism is a stage in the process of the loss of faith in personal Providence. Eighteenth-century Deism still awards Providence a place in the good order of nature, but no longer in a personal way. As Charles Taylor remarks in *Sources of the Self: The Making of Modern Identity* (Cambridge: Cambridge University Press, 1989): "The Deists had no more place for the 'particular providence,' God's interventions in the stories of individuals and nations, which were at the center of much popular piety and were extremely important to the orthodox" (p. 272).

22. See Hartmann, *Reproductive Rights and Wrongs*, pp. 271–96.

23. Each to his own boat, lest someone is picked up who sinks all. Cf. Paehlke, *Environmentalism*, pp. 41–75.

24. The concept of 'passing through and replacement' associates 'giving birth' with 'dying off,' in the sense of long-term continuing processes.

25. See Elizabeth Potter, "Gender and Epistemic Negotiation," in Linda Alcoff and Elizabeth Potter, *Feminist Epistemologies* (New York: Routledge, 1993), pp. 161–86. In her original essay, Potter considers all epistemology as part of social

negotiations: "giving a conceptual analysis, a phenomenology, an archeology, an account, a philosophy, or a theory can be part of a negotiation among many parties over what will be accepted as authoritative knowledge" (p. 170).

26. See Ruether, *Gaia and God*, pp. 85–8, on *metanoia* and catastrophe thinking.
27. Concerning this point and the critic's projection beyond the frontiers of one's own sphere, see Hans D. van Hoogstraten, "Europe as Heritage: Christian Occident or Divided Continent?" in *Bonhoeffer's Ethics: Old Europe and New Frontiers*, edited by G. Carter et al. (Kampen, the Netherlands: Kok Pharos, 1991), pp. 97–111, especially 104–7.
28. See Lothar Kettenacker's instructive essay "Der Mythos vom Reich," in Karl Heinz Bohrer, *Mythos und Moderne: Begriff und Bild einer Rekonstruktion* (Frankfurt: Suhrkamp, 1983), pp. 261–89.
29. See Ernst Bloch, *Das Prinzip Hoffnung*, 3 vols. (Frankfurt: Suhrkamp, 1959).
30. See Barbour, *Ethics in an Age of Technology*, pp. 3–10, 26–33. Barbour uses Maslow's five levels of need, which are helpful to understand humans' needy position. The Austrian philosophical anthropologist Arnold Gehlen typified the human being as *Mängelwesen* (deficient being) in his book *Der Mensch: Seine Natur und seine Stellung in der Welt* (1940; Frankfurt am Main: Klostermann, 1993).
31. See Marshall Berman, *All that Is Solid Melts into Air: The Experience of Modernity* (New York: Penguin Books, 1988), pp. 37–86.
32. See Berman, *All that Is Solid*, pp. 87–129.
33. A special continuing existence is given to the martyr, both in the community and with his Heavenly Lord. This is described (in Dutch) by C. W. Mönnich in his book *Koningsvanen: Latijns-christelijke poëzie tussen Oudheid en Middeleeuwen 300–600* (Baarn: Ambo, 1990), pp. 343–83.
34. Marx defines this as the change of money into capital, indicating a process of transformation, *chrematistics*. See Karl Marx, *Das Kapital: Kritik der politischen Ökonomie* (Vol. 1, 1867; Berlin: Dietz Verlag, 1973), pp. 161–91.
35. See Franz Hinkelammert, "The Economic Roots of Idolatry: Entrepreneurial Metaphysics," in *The Idols of Death and the God of Life: A Theology*, edited by Pablo Richard et al. (New York: Maryknoll, 1983), pp. 165–93, especially 165–8.
36. One of the most recent ideologies to employ this metaphor was the Nazi doctrine espousing that Jews were sickness-provoking cells in the healthy body of German society. This example shows the hazardous applications of this metaphor. Based on her book, *The Body of God: An Ecological Theology* (Minneapolis: Fortress Press, 1993), Sally McFague doesn't seem to be fully aware of these associations. She mentions fascism and Nazi ideology, but she states only that "elements can also be seen in some forms of communism and, in fact, in any from of societal organization that works from the top down rather than the bottom up" (p. 36; see also p. 225, n. 3). She neglects the ideological aspect: People believed the organic model to be an exclusive and an excluding model.
37. Hartmann, *Reproductive Rights and Wrongs*.
38. This is Marx's expression to indicate capital's absolute and limitless gluttony.
39. There is a parallel with the opinion about fate and fatalism, which I discussed in this chapter's first section.
40. See the encyclical *Centesimus Annus* (1992), which combines, as encyclicals used to do, moral condemnation of economic egoism and moral endorsement of family values.
41. In welfare work, attention is given to the victims of 'sinful relations.' The preceding

victimizing processes, however, remain in the dark. Again, the parallels between the conception of money and the conception of life are striking here.

42. See Dietrich Bonhoeffer on Luther's doctrine of offices and mandates in *Ethics*
(London: Fontana Library, 1964), especially pp. 196–213, 263–302.

43. See Susan Griffin, *Woman and Nature: The Roaring Inside Her* (New York:
Harper and Row, 1978).

44. Hartmann, on the contrary, speaks of reproductive rights as an important field
of human rights; see above.

45. Abraham did this in prehistoric times; see this chapter's first section.

46. Oedipus's father, able to see his son only as a threat to his own existence,
didn't get this far; see this chapter's first section.

47. Having explained this concept earlier, I repeat the argument here. Hans-Georg
Gadamer writes about the waves of enlightenment in "Mythos und Wissenschaft,"
in *Christlicher Glaube in Moderner Gesellschaft*, Vol. 2 (Freiburg: Herder, 1981),
pp. 6–42. The second wave of enlightenment occurs as late as the eighteenth
century and "reached its climax in the rationalism of the French Revolution's
era." Gadamer mentions a third wave "in our century's movement of enlightenment,
which, for the time being, reached its top in the atheism's religion and its
institutional foundation in modern atheistic orders of the state" (pp. 8–9).
Considering our century as a continuation of the eighteenth century's Enlightenment,
I would prefer to rewrite Gadamer's latter characterization of a third wave of
enlightenment as a development that is also critical to the economic-technological
system we live in.

48. See, among others, Harvey Cox, *The Secular City: Secularization and Urbanization
in Theological Perspective* (London: SCM Press Ltd., 1965), and Arend Th. van
Leeuwen, *Christianity in World History: The Meeting of the Faith of East and
West* (London: Edinburgh House Press, 1964). These positive views toward
progression in the 1960s seem rather naive now.

49. The German language uses the unique expression *Aufheben*, which means
overcoming and at the same time bringing to a higher level. Cf. especially Hegel's
dialectical method.

50. The Frankfurt School did much research in this field; see Martin Jay, *The Dialectical
Imagination* (Boston: Little, Brown, 1973), especially chap. 2.

51. See Harvey Cox, *The Seduction of the Spirit* (New York: Simon and Schuster,
1973).

52. See, among many others, Fritjov Capra, *The Turning Point: Science, Society
and the Rising Culture* (New York: Simon and Schuster, 1982). Capra discusses
economic and political structures, but not in the deep sense of this book.

53. Social contract theoreticians such as Hobbes and Rousseau had different views
in the philosophical and political spheres concerning human beings. See Joshua
Mitchell, *Not by Reason Alone: Religion, History, and Identity in Early Modern
Political Thought* (Chicago: University of Chicago Press, 1993).

54. Important here are Marx's debate with the Hegelian tradition and his efforts to
overcome Hegel (Hegel *auf zu heben*). See Karl Löwith's classic study *Von
Hegel zu Nietzsche: Der revolutionaire Bruch im Denken des neunzehnten
Jahrhunderts* (1941; Hamburg: Felix Meiner Verlag, 1978).

55. As we learned from studying Adam Smith's moral theory–ideology (see chapter 1),
the modern individual is a collective person, bearing the universal law—the
'impartial spectator'—in the self.

56. Ludwig Nagl deals with this point in *Habermas and Derrida on Reflexivity* in

Enlightenments: Encounters Between Critical Theory and Contemporary French Thought (Kampen, the Netherlands: Kok Pharos, 1993), pp. 61–76. See also Hans D. van Hoogstraten, "Ethics and the Problem of Metaphysics," in *Theology and the Practice of Responsibility*, edited by Wayne Whitson Floyd Jr. and Charles Marsh (Valley Forge, PA: Trinity Press International, 1994), pp. 223–37.

57. In Europe, moral subjects are forming groups that use money in an alternative way. For example, advocating a new tax system that they call Ecotax plus, they try to avoid the mistakes of nineteenth-century utopian communities. In doing so, they behave as political and economic pressure groups. See also this book's epilogue.

ENVIRONMENTAL ETHICS:
DIVISION AND CONNECTION

AESTHETICS AND POLLUTION

THIS CHAPTER OFFERS INSIGHT into the deconstruction of enlightenment through a focus on the unusual combination of pollution and aesthetics. 'Pollution,' as a collective notion for all possible forms of contaminating and poisoning air, water, and soil, is definitely the most radical aspect of modern environmental problems. In addition to factual ongoing pollution, there is a spirit of pollution that causes injury and destruction. This spirit is the functional context wherein humans stand in a specific (spiritual) way in relation to nature.

Ongoing pollution is partly the result of our aesthetic attitude toward life. In the past few centuries, the aesthetic worldview has deeply influenced Western culture.[1] This aesthetic attitude toward life is intimately connected to ethics. Within the cultural condition of modernity, this aesthetic attitude has impacted norms, values, and rules in a dominant way. Psychologically, we practice an aesthetic determination of value.

To comprehend this, it may prove useful to stress that we are dealing with an honorary tradition wherein the aesthetic value of things was considered identical to the thing's intrinsic value. The created reality had beauty in itself; hence, people were called to worship. Both in nature and in culture, 'creations' always pointed at a creator who instilled beauty and sublimity in his creation. A well-educated person was expected to show reverence for the sublime. Appreciation for the sublime demonstrated cultural correctness.[2]

Another tradition originated in our culture approximately one hundred years ago, and it is more widespread: the tradition of the utilitarian-instrumental-economic value. Originally, these were separate traditions, but during the twentieth century, they gradually became firmly connected.[3] In China, for example, it was quite normal to connect aesthetics to the daily use of things; in the West, however, this connection indicated a breakthrough in established sociological and psychological patterns.[4] Indeed, art became a common good, and commerce became closely linked to aesthetics. The acknowledgment of creation-based value was replaced by the granting of value to objects and products. Desirable objects acquired aesthetic value, and the preferred acquired a special meaning. Need became preference.[5] The Western imagination connected the beautiful with the desirable.[6] People gradually came to believe that beauty was the basis of a happy life: One had to possess both physical beauty and the 'aesthetically right' things. We can observe the process by which the individual acquires respect for style and taste. The individual earns aesthetic respect by adopting the accepted standards of a particular social circle.

The twentieth century has witnessed the development of an unequaled ideological side of modern society. The development is actually a shift: What was once reserved for the elite now becomes common property. Mass media and technology cash in on this development. The mass media use the desirable to manipulate the public.[7] Hence, the value is popularized via the fusion of commerce and aesthetics. Beauty is desirable, and the object of desire is beautiful. Technology connects with design, which is sensitive to trends; again and again, a generally acceptable fashion arises on various fields. Technology successfully feeds on and into the 'generalized' taste. Commercials recommend the goods as something grand; sometimes the product is literally gigantic, and other times the recommendation concentrates on one sublime characteristic. In either case, however, the product is guaranteed to enlarge the user and owner. The admiration of family and friends validates the guarantee. We don't have to look far for trivial examples: houses, cars, watches, clothes, technological equipment, vacation packages, and so forth.

The aesthetic attitude toward life relishes the external, the appearance, the product.[8] Once, the beautiful represented the sublime (metaphysics); now the beautiful is functional. The symbolic value that once connected beauty with status and honor has become an exchange value that indicates wealth and well-being. The 'right' product must be visible. All attention is focused on what can be done with the exhibited, in which environment it should be used, and which behavior will make it function. This can be turned around: The user and owner function best in this way, as opposed to any other way. Here, under the flag of modern aesthetics, we meet the effects of the fusion between the human and the material object. Human relationships are profoundly influenced by this fusion. Quality of life becomes based on new products, fashion, and appropriate acquisition in general. Quality of life becomes synonymous with the means to acquire desirable objects, technology, and specific aesthetics. A characteristic of this attitude toward life is that people choose what can be exhibited. Aesthetic choice is imposed on the modern individual; it no longer arises from within. Thus, beauty is no longer a quality or an inner feeling of aesthetic appreciation; instead, beauty has become a thing that can be seen, touched, and possessed.

We can also discern the psychological effects of our economically exploited aesthetic tradition. It is typical of aesthetics to deny that which it does not want to see, such as poverty, disease, age, and death. People are inclined to acknowledge only the pleasant side of life, and they connect themselves only to this side. The dark, mean, dirty side is not acknowledged. If they can, they deny its existence and reality altogether.[9]

This denial significantly impacts the domains of politics and economics. In the age of sovereigns, the king was regarded as God's representative. Hence, whoever's back bore the burden was ultimately irrelevant. In an economically controlled culture, the product is exhibited, recommended, and traded. People know how to directly participate in this process. But they do not know how to address the fact that the consequences of production and consumption are social and ecological problems.[10] Most people prefer to deny, ignore, and forget these consequences.

This issue is grounded in psychology. We learn to concentrate on what goes in, not on what comes out. We all leave the anal phase behind, so to speak. A fixation on excrement is unhealthy. Excrement is flushed away, and it has nothing more to do with us, nor we with it. It is a matter of hygiene. However, this is not the way it has always been. Humans lived with dirt, putrefaction, and plague fumes on a daily basis for centuries.[11] Thus they must have known that production and life cause waste and death. This knowledge must have influenced the spirit and the way in which humans connected with nature. We know that wealthy people employed others to dispose of their waste, so that they could live in the imagination of beauty. In our own time, waste is kept away by sewers, refuse dumps, and (almost) invisible exhaust fumes; factories are placed far out of sight, so that they can be kept comfortably out of mind.[12] The power of this aesthetic attitude toward life is such that we still have not causally connected it to its polluting consequences. Moreover, even though waste and pollution afflict nations that are unable to provide for people's primary necessities, the rich countries still have not connected this to their own actions, policies, and preferences. As a consequence of the present world economic situation, wealthy nations are quite content to let the poorer nations perish, so long as their demise is unseen and unheard. The ostrich approach, however, cannot erase the fact that our aesthetic attitude is tied to the world's deep economic relations.[13]

Our aesthetic fascination strengthens our perpetual existential refusal to make a real change. A person who is fascinated by the emanation of desirable objects wants immediate contact, an instantaneous relationship. The aim is to own and use the object of desire. It is not the cost of the thing or its history, the process of its growth and production, that fascinates the person; instead, it is the present appearance of the thing itself. The thing will be replaced in the near future when it no longer works or when it becomes obsolete. The fixation on the present also means an ongoing preference for the new. Since this fascination is already within the subject, business uses it to manipulate the subject. By manipulating and satisfying the preference, the accompanying phenomena, such as pollution and exploitation, are kept out of focus.[14]

Our aesthetic attitude toward life is continually encouraged by our mass producing, consuming, media-saturated society. In our global trade and transport system, the fascinating object can come from literally anywhere. Provided there is enough money, the possessing potential is unlimited. This means that everyone eventually becomes fascinated and proceeds to connect with the offered, on the imaginative level at the very least. The desire for 'closer contact' and for possession has captured the world's imagination.

This intensifies and narrows the human's spiritual capacity to form relationships (intensity can increase through narrowing). This means that people are both unwilling and unable to see anything that is not in the spotlight. Values are narrowed to the freedom to buy and sell, use and enjoy. Norms are adjusted to this rigorous compass. Rules whereby the aesthetic relations can show results need to be outlined carefully.

The aesthetic attitude toward life does not know how to handle the pollution that results from economic, technical, and political actions. It does not know how to ascribe meaning to it. The waste disappears—this is our direct experience. Yet in the last few decades, it has become quite clear that the invisible waste has reappeared with a vengeance. But we typically experience this indirectly; indeed, we rarely meet with the direct consequences of our polluting practices. The people who experience the direct consequences are the inhabitants of the Third World. For this reason, ethical reflection must be addressed to the First World. We must avoid running with the hare and hunting with the hounds of Western environmental thinking, those who, in the name of environmental considerations, want to keep the Third World down.[15] The Third World directly experiences the First World's pollution.

DEBT, GUILT, AND INSTALLMENT

The desire to possess and to connect oneself to the desired object is overwhelming in modern society. To satisfy this desire, money must be available. I have already discussed the specific way that banks provide the necessary money. And I have identified money as the circulating blood in the economic body.[16] I now examine the psychological and sociological aspects of the way in which money functions. In doing so, I show how structural power operates, in the hope that a deep understanding of these functions and operations will empower us to contest them.

We are on the fields of morals and psychology when we acknowledge the debt-guilt relationship in modern economy.[17] In this context, guilt means financial debt rather than judicial, religious, or moral guilt. But it will become clear that economic practices simply cannot exist in a sphere separated from judicial, religious, and moral practice and guilt.[18] In the discussion of economic debts, we must remember money's three functions: medium of exchange, means to measure value, and means of capital accumulation.[19]

The disposal or possession of a negative balance is appropriate to our modern money economy. We need to lend and borrow money. People enter into relationships with one another on the basis of money. All three functions of money are present in these relationships. In fact, if one of the functions, or meanings, of money is not present, the entire relational framework immediately becomes unclear. We can recognize the presence of money's three functions in ordinary transactions. The applicant for a loan needs money as a medium of exchange. The supplier of the money needs applicants in order to receive interest. Interest must be received to meet inflation and to continue the accumulation of capital. The applicant for a loan must be able to prove that he or she is solvent, and the person's value is measured by his or her probable income. The loaner decides whether or not the applicant is likely to bring in the interest.

Over the last few decades, procuring credit and incurring debts have become commonplace, on both micro and macro levels. The astounding coherence between credit and debit on the micro–macro scale has led to two crippling developments. First, wealthy nations are strangling poorer countries with gigantic encumbrances. And second, the average wealthy-nation citizen must participate in maintaining the growth process, even though it is destructive toward nature and humanity.

Extending credit is a daily business. Creditworthiness is measured in various ways; for example, the creditor takes the applicant's future prospects into account. Credit is frequently offered to students, because they can be expected to earn a decent wage in the near future. People with few prospects are rarely offered credit. Thus, even in this early stage of furnishing credit, judgment is passed and power is executed, power *in actu*.[20] The applicant accepts these terms, because he or she is in a dependent position: The applicant needs the money. The power is activated the moment the general feeling of dependence becomes concrete in relation to the 'giving institution.' The lender will do anything to meet the demanded obligations for corporate survival. For the lendee, the immediate need of collecting the necessary money outweighs the later need to gather the financial means for repayment and interest.[21]

The creditor and debtor play a refined economic game. Providing credit increases influence and purchase power. Creditors invest their money when conditions are suitable and financially attractive. In order to enlarge their capital, creditors require applicants to demonstrate present and future solvency. The government promotes the lending and borrowing of money by making interest fiscally deductible. This economic game enables people to live a rich lifestyle and to possess many things, but all is procured with borrowed funds. People are free to act as they please. Money as a medium of exchange is at their disposal as long as they agree to supply the interest that must be generated from the originally offered sum (money as a medium for accumulation). This process can continue as long as the person's income continues to guarantee creditworthiness (money as a value measurer).

This new wealth of 'negative possession' has significantly influenced behavior and psychology. Because people live on credit, they do not consider themselves wealthy. This is also true for countries; consider the prominent budget deficit discussions on both macro and micro levels. The credit card is a perfect symbol of our system's underlying absurdity: People in relatively poor financial situations have wealthy lifestyles. The catch, however, is that the distribution of goods and services reaches only the people and the countries that can guarantee solvency.

Economics and politics cooperate in the organization of the social system. This simply means that government supplies the facilities and conditions for the economic game. The system tries to procure credit for as many as possible, but in order to do this, the debtor must have the opportunity and the ability to meet the creditor's demands. The government assists the system by offering a lower tax on 'negative property,' meaning financial debt. This encourages and helps people to make purchases with borrowed funds. Debt also stimulates people to earn money, and this, in turn, stimulates the entire process of the growth of money.

There is a command structure in the free-market economy that determines power positions. The established way is the only way in which the economy can execute its task of providing for our needs in a 'providential' manner. From an ideological standpoint, debtors and creditors occupy strategic positions that must be supported and defended by political and judicial means. The morality and righteousness of the system are incontestable. The power is not arbitrary but is accurately determined, meaning that it is experienced as necessary and is therefore deemed just.[22] This ideology is hardly palatable when we take the other's experience of our practices into account.

Western power structures and ideology have been established in the Third World. Third World countries do not have the means to procure the creditor's interest. Often, new interest accumulates on top of unpaid interest, and the possibility for repayment becomes nonexistent. The disadvantaged country is now in a position of proving eventual solvency in order to attract desperately needed creditors. The Third World country resorts to swearing undying allegiance to the First World's power structures and supporting ideology. The creditor often sees this as sufficient solvency for its own purposes.

Once the poor nation declares its loyalty to the laws of the economic system, the creditor establishes itself in the poorer country.[23] The poor country accepts the employment of ecologically destructive and exhausting activities in order to try to stay alive. The destruction of the rain forests is the most well-known example. The power position that nations and banks hold *together* is of the utmost importance. Both the International Monetary Fund and the World Bank are governed by the rich countries, where capital accumulates. The rich countries disperse the capital throughout the world, but it ultimately returns to the hands of the rich as surplus value.

We must recognize our compulsive political and economic acting so that we can react as decision-making beings. Contemplating our compulsive power structures generates insight into the destructive character of our current socioeconomic system. And when we integrate environmental ethics, which is gaining popular acceptance, we begin to form the base for an enlightened community willing to implement change. The critical character of environmental ethics is established when two conditions are met: First, people must want to change, and second, economic guilt cannot be separated from judicial, moral, and religious guilt. Concerning the former condition, the will to change is a matter of the power of the spirit. As Nietzsche discusses, this spirit locates people in the active, positive, confirming, and productive power field, as opposed to the reactive, negative, denying, and consumptive power field. The latter field conquers by division. It is dualistic, because it feeds itself with resentment and bad conscience: Guilt engulfs and divides.[24] Separating economic guilt from other types of guilt perpetuates dualism and generates the consumptive, negative power field. And as long as we continue the separation, we are socially and psychologically bound to this power field.

To demonstrate this, we can begin with 'the payment.' The lending person expects repayment; the borrowing person must repay. These two persons are contractually connected, and they are mutually dependent on each other. The creditor depends on the will and abilities of the one who owes money. If the creditor is smart, the risk will be spread so that dependence is relative. The situation is different for the debtor. Failure to pay means that the debtor will be 'stripped' and will be unable to receive new loans.[25] The debtor must pay. The debtor's obligations are judicial matters, and, morally speaking, the debtor lives with the conditions and consequences of debt. Debtors picture themselves as people who need to settle a debt.

This final consideration should be a self-evident one, but often this is not the case. We live with the knowledge that we are free, and guilt disrupts this freedom. Guilt and redemption are widespread religious phenomena. In the Christian tradition, the guilt of the sinner is redeemed by Christ's reconciliation. The philosopher Hegel thought that the Enlightenment and the recognition of human rights largely extinguished the problem of guilt.[26] However, it seems that humans have a psychological mechanism for guilt, and guilt is tied to certain relations and to the knowledge that one has incurred guilt by some action. We must understand this guilt mechanism in order to establish a critical environmental ethics. Ironically, the problem of guilt is liberation: You are free only as long as you pay. This holds true in various fields. In the *religious* field, a mediator pays. In Christianity, Christ mediates, but many religions have a similar type of guilt and payment system.[27] In the moral sphere, human acting is crucial: 'I did what I could' means that the obligation was met and the other's rights were attended to as fully as possible. On the judicial field, we have sanctions and punishment, often expressed in monetary terms.

In the sphere of economy, nature pays for human guilt. In modern society, guilt has a monetary character, so in order to repay, we need to create money. *In abstracto,* the money has been created already: One's money is with the other, and the other lives on credit. The amount varies from minimal to exorbitant. Ironically, the United States is both a major creditor and the world's greatest debtor. This is possible because guilt concerns abstract agreements in which the monetary amounts are merely symbols, at least until one must pay. Then heaven and earth must be moved—literally—to let nature fulfill its 'reconciling' task: The implicit value of nature must be made explicit; nature must be cashed in. Rich nations have a cash-in advantage over poorer nations. This is due to the technical means to plunder and exploit, means by which a country becomes creditworthy and thus relatively rich. Only the readiness to exploit humans and nature will provide enduring solvency.

This brings us to the sustainable development of the economy. Only if there is payment—if the rich creditor gets richer, if capital accumulates—can the economic game continue. This includes continuing the well-known consequences of these actions, on the basis of the rights aligned with power positions. The obligation to pay shackles people. Even if they wanted to act differently, they could not, because it would diminish their solvency. They buy their freedom by meeting the latest economic truth. They themselves are mediators for the growth of the money suppliers' capital; this is true even if the money appears to be made by 'development cooperation.' It is not even cynical to state that the 'cooperation' certifies economic 'development'; it is an obvious fact.

I have pointed out the fundamental aspects of the power constraint and the occupied power positions therein. Yet two more elements need to be included: nature and labor. The parties that want to receive credit must show that they can provide profitable labor. They must be willing to do the work, even if the cooperation involves something that they find morally reprehensible. They must do this in order to sustain economic development. When an effective labor force declines, when there is no employment, when nature is used up, and when technological development decreases the need for labor and raw materials, the cash flow stops; then a hopeless situation develops. The Third World, closely followed by the Second World, feels these developments in alarming ways. The First World knows that the consequences of its economic thinking and acting will hit home too, sooner or later. We can start overturning these developments by assimilating and communicating a critical ethics of guilt relations. We must define the ambiguous morality and psychology that maintain the creditor-debtor relationship. After doing this, we can begin developing healthier attitudes and policies that will wipe our short-term, shortsighted interests off the map. Then we would have designed a moral and practical framework for the new paradigm of the third wave of enlightenment.

RIGHTS OF HUMANS AND OTHER SPECIES

The issue of rights is important to our ethical-environmental discussion. The establishment of rights could protect life that is currently being threatened by various activities. In spite of popular talk, animals, natural objects, and future human generations are entirely without rights in our society. Before we discuss the (un)desirability and the (im)possibility of the granting of rights to different categories, we must review the rights that humans refer to and claim. In the Anglo-Saxon rights discussion, a distinction is often made between legal rights and moral rights:

> In general, a right is an individual's entitlement to something. A person has a right when that person is entitled to act in a certain way or is entitled to have others act in a certain way toward him or her. The entitlement may derive from a *legal* system that permits or empowers the person to act in a specified way or that requires others to act in certain ways toward that person; the entitlement is then called a 'legal right.' . . . Legal rights are limited, of course, to the particular jurisdiction within which the legal system is in force.
>   Entitlement can also derive from a system of moral standards independently of any particular legal system. . . . Such rights, which are called 'moral rights' or 'human rights,' are based on moral norms and principles that specify that all human beings are permitted or empowered to do something or to have something done for them. Moral rights, unlike legal rights, are usually thought of as being universal insofar as they are rights that all human beings of every nationality possess to an equal extent simply by virtue of being human beings. [28]

This is a clear and helpful description from a well-known textbook on business ethics. However, it does not (and cannot) grasp the many problems in this tricky field. It is revealing to examine which human and moral rights are considered relevant in the judicial and political spheres. People have the right to appeal their judicial positions; thus legal rights can be examined. But although we can discuss the interpretation of these rights, we cannot discuss the fact that people have them.[29]

In the allocation of moral rights, people decide whether or not other humans (living now or in the future) and other living creatures have certain definable rights. To prevent an absurd misconception, we must realize that only people in the present can claim rights.[30] In all other circumstances, we can speak only of allocated interests. Strictly speaking, we should understand the interests of these others: animals, future generations, and natural entities. The motivation to advocate these 'rights' is based on solidarity and connectedness. In forming opinions, different people are led by different motives, norms, and values. But the reigning social ideology and worldview influence the decision-making process. Deep ecology, for example, establishes a metaphysical basis for speaking on moral rights; what we ought to do is derived from our being a species in the universe of species.

In everyday society, we see a clear pattern of priorities concerning people's rights and duties. Legal rights are by far the most important. They are firm and are stipulated by time and place. These are the rights that have been consolidated over the last few centuries for Western citizens. Because they are connected to economic and political relations, they are ironclad. They, for example, make it possible to live on debt, and they protect the debtor and creditor alike. The allocation of moral rights coincides with norms and convictions. The transformation of moral rights into legal rights is quite a step. The discussion is often begun by those who already have a strong legal status. Hence, interests always play a part in the allocation of rights to others. If the allocation of rights is to be discussed on behalf of the other's interests, the central question, whether open or covert, conscious or not, will always be whether one's own position will remain strong. For this reason, rights discussions often turn into some sort of social battle. The other is considered the stranger and cannot have the same rights as the self. If we keep this in mind, we might be able to bypass the idealistic, abstract spheres in which discussions on rights often take place. The United Nations' Declarations on Human Rights are a large-scale example. As long as such declarations remain general and vague, everyone agrees with them. But when the declarations are applied in real places to real circumstances, disagreements and real problems arise.

The classification of legal and moral rights should not cause us to forget that their acquisition was a process. The now self-evident, fixed constitutions and laws are the result of historical development. In many instances, they are the outcome of fights between interest groups in the areas of race, sex, and class. In the deep economic sphere, we can trace the origin and fixation of legal rights that were deemed necessary for the system's development. In order to legitimize these legal rights, they first had to be accepted as general human rights. John Locke (1632–1704) fiercely advocated the right to private property. By arguing in a typical Western way, he formulated the generally felt desire for a specific human right: the right to live. The bottom line is that every human has the right to live, and as one needs provisions in order to survive, one has a right to them as well. People can provide for themselves through their own energies, because the body is the property of the person.

> Though the earth and all inferior creatures be common to all men, yet every man has a property in his own person; this nobody has any right to but himself. The labor of his body and the work of his hands, we may say, are properly his. Whatsoever then he removes out of the state that nature has provided and left it in, he has mixed his labor with, and joined to it something that is his own, and thereby makes it his property. It being by him removed from the common state nature has placed it in, it has by this labor something annexed to it that excludes the common right of other men. For this labor being the unquestionable property of the laborer, no man but he can have a right to what that is once joined to, at least where there is enough and as good left in common for others.[31]

In this notorious part of *The Second Treatise of Government*, Locke stresses only the proprietary right. He can still take for granted that each human has the disposal of his or her own working power, and that the individual has the proprietary right by 'mixing' his or her labor with nature. This is all based on the right to live. Locke stresses the proprietary right in connection with the use and availability of the soil or nature. Property and labor belong together. We can clarify this by analyzing the ideology that accompanied the colonization of America: The Indians did not work and therefore had no right to the land.

In the liberal tradition, of which Locke is a main representative, the proprietary right is the most important right. In some cases, always depending on position, the proprietary right was more important than the right to live. This late-seventeenth-century theory deeply influenced the succeeding centuries' development of Western rights theories and practices, including social contracts.

Locke's ideas were a progressive landmark, in that they opposed the common law of aristocratic landowning families. As such, Lockean ideas mark the end of one era and forge the beginning of the next, in which the ownership of the means of production and its profits is safeguarded as a legal right.[32] We could say that the proprietary right has become the mainstay of society. All other rights originate from it. The mediating role of money is foundational, because the judicial right to property and inheritance is economic in our age.

It is impossible to separate the economic factor from psychology and aesthetics. In discussing rights in connection with environmental ethics, it is essential to consider the objective, economic meaning in which all subjective allocations of meanings are embedded. This becomes clear as soon as we look at expropriation. This can take place only on economic-political grounds (roads, development plans, and so forth). In the name of the *bonum commune*—the 'common good' and the 'general interest'—the most precious rights can be violated if the violation serves a higher (economic) goal. Judicial rights must fit into the social system.[33] According to Locke, however, the central proprietary right, from which stems all exploitation, pollution, and unjust division, originated in the control that people have over themselves. I have a right to my own life and body, therefore my life and body are inviolable. This human right, considered universal by many, means that my singularity originates in what I do not share with others. My singularity is my authenticity.

Does this mean that I can assert my right to property regardless of what follows my act of assertion? The English words 'property' and 'proprietary' are compositions of the word 'proper.' This stems from the Latin word *proprius*: the unique thing about someone, exclusive to, that which one does not have in common with others. The term 'proper' stands for both this intrinsic characteristic and a moral qualification: honest, admirable, and worthy. These exact qualities, as Adam Smith told us, are the moral and functional basis for the commercial society's success. As we can see, rights and morality are closely connected in deep economy.

But when we focus on the sphere of (deep) ecology, the contradiction between moral rights and the economy is evident. The 'rights' of nonhuman species and natural objects are related to life in the biosphere and to intrinsic value. The rights of future generations are related to the earth's long-term life-giving capacity. Presently living humans are challenged to act in a morally responsible way by acknowledging the other living beings' position and conditions. The whole environmental rights discussion is dictated by the natural order. It petitions humans to know their place and to stop their destructive activities. Is the environmental rights discussion a fruitful and effective development? The answer depends on the character of the discussion. If we consider the development of rights as a historical process, and not just as the awarding of metaphysical rights fixed in an ontological order, the discussion is headed in an effective direction. This is demonstrated by the following examples.

First, it can be argued that the usual connection between the right to live and the right to property is invalid, because the proprietary right in practice damages the right to live. This realization forces the critical contemplation of the position of the rights of the subject. The balance between life and death is at stake.

Second, animal rights activists presuppose the right of several species to live in freedom. All these species have a position in the ecosystem, specific needs, and behavioral codes. Thus, according to their argument, we need to honor a connection between freedom and natural conditions. We should realize, however, that nature has always been used in ideological ways within the cultural and religious history of freedom and slavery. Thus, we need to be wary of a dangerous unilateral overstressing of nature in connection with freedom.[34]

Third, we have to reckon with the rights of future generations. This means a firm emphasis on the duties of those living now, concerning the possibilities for the ones who will live in the future. At the same time, we can critically reconsider the overemphasizing of rights for the present. (In the law of inheritance, there is a recognition of the rights of future generations, although it concerns family rights.)

Fourth, we should realize that the plea for moral rights takes place within a discussion among people. Thus, the important point is how people are connected to themselves, to one another, and to nature. The conceptions that arise in this position-bound discussion are profoundly influenced by interests. Since moral rights usually have a universal character, they potentially clash with specific personal rights. Allocating moral rights is not a neutral act. If I allocate a moral right to someone else, then I have a moral duty toward the subject who has the moral right. And this means that I may have to waive my own rights; I may have to consider my interests inferior to the interests of those who are hurt—even indirectly—by my pursuit of my personal interests.

There is one important advantage in discussing the rights of subjects who cannot speak for themselves: It undermines a human interest–based interpretation of the doctrine of natural rights. All the 'social contract' proponents advocate the primacy of 'by nature' rights, and this is the source of much argumentation and theory.[35] In fact, it expresses Western dualistic thinking in an exemplary way. We first speak of obtaining universal rights, and in the next breath, we allocate different rights to different groups of people. We then reason that since humans 'naturally' differ, they can have different rights. This ideology is easily applied to sexes, races, and classes. The issue of needs is customarily incorporated into this discussion. The contemporary rights discussion reconsiders the relation between nature and history. It recognizes that people 'make' history by interpreting natural laws and exploiting natural energies. Does this mean that nature has become dependent on human history? We can answer this by connecting the supposed dependency relation with the theme of sustainable development.

SUSTAINABLE DEVELOPMENT: FROM HISTORY TO NATURE

The issue of sustainable development is the focal point of environmental ethical discussion. The need for sustainable development is evident to all environmentally aware persons. Historically, few ideals have so rapidly succeeded in causing such a great amount of frustration. Although everyone agrees on the absolute necessity of a formulated goal, it seems impossible for everyone to agree on the formulation. The Brundtland Report vigorously pleads for sustainable development but simultaneously wants to maintain economic growth.[36] Most analyses neglect the causal coherence between nonsustainable development and economic growth. A profound discussion arose in the wake of the Brundtland Report concerning the very conception of the term.[37] The notion of 'sustainable development' indicates a continuous progression: Sustainable means 'maintainable' or 'procurable.' Thus the present development must be maintained. In the Brundtland Report, the tension between two separate interests is apparent: the wish to maintain current economic efforts to preserve society, and the demand that the environment bear the consequential assaults of these economic efforts.

The goal is usually formulated as the sustainability and tenability of nature and the preservation of the ecological balance. But how to achieve this is an ongoing debate. The authors of the Brundtland Report could not afford to suggest economic shrinkage. Their conviction is that a lot of money will be needed to reach sustainable development, and they assume that the money will have to be made in the proven way.[38] Those who object to this notion have not supplied alternatives for financing development. I suspect that the underlying dualism will remain hidden as long as we stress the sustainability and maintainability of nature only. In our type of society, the focus is on progression, and there is no indication that this firmly established focus will be waning in the near future. In fact, there is evidence to the contrary. Since the collapse of communism, liberal capitalism has been spreading at a rapid rate.[39] Sustainability, as economists discuss it, means that all obstacles for *this* development need to be overcome. In accordance with this reasoning, the perpetual creation of money creates the conditions for sustainability, and sustainable economic development creates the demand for money.

Free-market economy's religious character is deceptively present at every turn. Economy and theology meet, linguistically and conceptually, in 'providence.' Modern humans are dependent on economic acting to 'provide' the means to diminish pollution and resource depletion. The word is useful in a theological sense. Indeed, the theological content sharpens the intrinsic irony of a providence without transcendence. Just as God's providence can be regarded as the crux of the Christian belief, the trust in the economy and its underlying theory of durable prosperity is regarded *conditio sine qua non* as assurance of maintaining our society and the individual's position within it.

Characteristic of our godless age is that people no longer depend on the incomprehensible providence of an external institution. Instead, people provide for themselves with rationality and a willingness to play the system's game. And they continually try to abolish the difference between manipulative and nonmanipulative social relationships. This economic game is played on various fields. This is the essence of the aesthetic-economic attitude toward life, in which people adopt the behavior and character of the rich aesthete.[40] With imagination and the guarding of rights, society appears to be able to keep its balance.

I have not strayed from the meaning of 'sustainable development.' This is all applicable to our desire to keep developing in a sustainable and durable way and in a way that leaves our environment in tact. However, some believe that these two desires are incompatible and inherently exclusive. If so, our often discussed and pursued goal would be a contradiction in terms, an ironic proposition indeed.

We must not let this ironic proposition develop into cynical indifference. The notion of irony implies that we are not in the sphere of determinism. Rather, we should think in terms of warning. Exploitation amplifies irony's warning. I define 'exploitation' as the extraction of energy from something or someone in such a way that prevents recovery. Exploitation seems to be the reason that nature is not sustainable. The warning is the boomerang effect: The originator eventually becomes the victim of his or her own practices.

The boomerang is aimed at the next generation. Our actions say that coming generations are responsible for their own business, yet coping with our business will be their environmental problem. Anyone who tries to see past this shortsightedness will realize that the plea for sustainable development opposes the present development. The pursuers of environmentally sound sustainable development are opting for a different social development.

Thus nature influences history in an unexpected way. People direct the course of nature's influence. The self-evident way in which Western civilization uses nature economically, politically, and ideologically is now being questioned. Increasingly, the connection falters between the naturally legitimizing and the material functions of nature. To comprehend the seriousness of this development, we must understand those crucial moments in Western history when nature offered sustainability to history. We know the sequence as being from history to nature.

Hugo Grotius (1583–1645) was one of the first great European thinkers who declared proprietary rights as natural rights in a logical way, explicitly combining them with God's providential will as creator. Largely due to his influence, the combination of civil rights and religion has become a solid part of Western culture. These 'naturally' endowed rights are the basis for the liberty to enlarge property by economically advantageous acting. 'Social nature' goes hand in hand with the providence of nature; it is the 'invisible hand' of divine Providence. The ideological exploitation of nature is confirmed by the possibilities of material exploitation. That is, we can exercise our natural rights and duties by using nature, as long as we do so responsibly. We know that we are acting in the right and responsible way when nature's 'cooperation' indicates as much. The ideology is that the harmonious functioning of property, production, trade, and consumption relations indicates that these things are naturally right.[41]

Western civilization's faith in nature's cooperative functioning is fundamental to our historical development and progression. In the Western conception, making history means to feel and to act independently of nature: to let nature work for you. Nature, the former lord over humans, is now subordinate to human freedom and progress. When the development appears to be unsustainable, it impairs our faith in self-evident truths, and a crisis develops. Insecurity regarding the legally anchored 'self-evident' and 'natural' rights arises from material obstacles.

We can postpone or deny the crisis for a long time, but the facts eventually start to speak for themselves. Natural species begin to disappear, and holes appear in the ozone layer. Suddenly, sustainable development is on everybody's mind—a signal of the encroaching definitive, critical moment.

Can such a crisis be fought and allayed in a technical way? Yes. If the urge is strong enough, we can find and use cleaner energy sources. But it will not be enough. We cannot change from demolition- and exhaustion-based development to sustainable development without a fundamental change in the relation between history and nature. And for this, we need to uncover the complicated relationship between subservience and authority.

The subservient role of nature must become one of authority. The difficulty we encounter is that nature has always been granted great meaning and authority. Humans feel dependent on what nature offers them. But this is the reason that they try to rule and control nature. The Western attitude is expressed in the specific way it connects with nature. As an aesthete, manager, and therapist, the human judges and chooses, rules and organizes, guards and heals, compromises if necessary, and is 'free of value' in ethics.[42] Nature holds a position of 'authorial subservience': While honoring its laws, we hold nature liable for our prosperity. This is the ultimate form of 'Smithonian' prudence.

This long development appears to be unsustainable after all. The efforts are too one-sided, and the ongoing process may bring final catastrophe. This one-sidedness arises from the subservient authority that is attributed to nature. As I stated, before, it is not enough to consider the process alone, thereby maintaining the relations between nature and history. We must think about nature in a fundamentally different way. The spiritual bond between humans and nature needs to be relieved of its apparent self-evidence. We must reverse the sequence to read: from nature to history.

I want to try to name this change, to catch it in words. There is a wind of change in society now; it can be felt here and there. A paradigm is a unity of theory and practice, and to date, we do not have a new one to replace our present destructive one. Those of us who want change have not established a new paradigm. The crisis is still not clear enough; we cannot feel its impact in our daily lives. So even the people who want change will not abandon the ruling paradigm until the forces of extreme crisis offer no other alternative. But by then it will be too late. The third wave of enlightenment must be born today.

NATURE AND HISTORY—WHICH SEQUENCE?

Can we provoke such a drastic social change that future generations will refer to it as a turn in history? Specifically, it would be a turn in the developing uniform Western history: the aggressive way in which people connect to and handle one another and nature, which ultimately nullifies cultural differences, because all act alike. Western history as an ideological concept of unity swallows typical cultural signs like a constantly deleting cursor. The engine of the Western way of making history is voracious and all-consuming. It stimulates the shaping of societies in ideological, political, and economic ways, to eventually include them in the well-functioning world system wherein dynamic relations bring forth greater productivity. We then speak of the dynamics of cooperation, competition, and exploitation. The question is whether it is possible, or even thinkable, to turn this Western-imperialistic way of making history in another direction.

The question appears too great and too theoretical to answer. Additionally, it apparently keeps us far from the urgent issue of what sustainable development actually means in the practical sense of production, consumption, and exploitation. Still, we must go in this direction to meaningfully discuss practical consequences. Environmental ethics supposes knowledge of the environment. In accordance with this notion, the environment *is* history.

A radical turnover is needed to stop our thinking in terms of contrarieties and thus prejudging the project of sustainable development as worthless. This turnover concerns a new order: nature → history. It also concerns a change in the present scientific models, including practice and theory. We need another paradigm. As nature suffers under the subservient authority we grant it, it becomes clear that we will have to endow nature with a different type of authority. The sustainability of modern history appears to damage ecological sustainability. It happens in such diverse and extensive ways that historical sustainability is, ironically, in grave danger.

The priority of ecological sustainability and the subordination of economic sustainability sound like a regression to premodern times. But I am not talking about a backward movement; on the contrary, I am discussing an environment-oriented development for the future. In this context, we should be aware of some basic facts: History develops, and every change takes place within its progress. Moreover, a new paradigm can appear only by using the elements of the existing paradigm that cannot be changed in any way. The change occurs gradually; elements of the old and the new coexist for a while. The coexistence, however, is not necessarily peaceful. In Europe, for example, religious wars had to be fought before freedom of expression and belief became a social fact. Centuries passed before democracy was accepted and organized, and it often developed next to forms of absolute state government. Scientific revolutions happened only after long discussions on which direction to pursue.[43] It is now up to us to rise to the difficult challenge of our own age.

In our era, the money economy is in charge. The 'war of all with all' has been transformed to a 'barter of all with all.'[44] Mentally, there have been many changes in the last few decades. At least now, many acknowledge the importance of a well-functioning nature—well functioning in cooperation with and judged from within a well-functioning human history.

Advocates of a radical change of paradigms will say that this means nothing, since we are still dealing with an anthropocentric vision. I think that this flat judgment is a cowardly excuse. We will remain anthropocentric in our stage of human development: Even when we reverse the order (nature → history), history will be qualified as human. The crucial point is the establishment of a different attitude toward how history is made, one that includes the organization of society. If history remains under the 'regime' of money and capital accumulation, nothing will change among people or between people and nature. The ongoing unsustainable development will proceed, leading to the death of many and, in the long run, to the death of all. The development is then irreversible.[45]

Reversible development focuses on mortal life and the life-death balance, not on the downfall of people and nature. It is under another regime altogether. In this regime, the ecological balance is in charge: The 'law' is the set of conditions for maintaining the natural system. A regime, in the sense of practicing government, can work only when the subjects accept its legitimacy. For us, this means changes in orientation, since we cannot possibly recognize two completely different regimes at the same time. A critical environmental ethics formulates the conditions that need to be fulfilled for change:

- It is necessary to define an independently acting and judging subject, in opposition to the current trend of denying the possible existence of such an individual or group.
- We can find arguments for this in the Jewish worldview, in which subjects who actively opposed the ruling trend were called into being (faith of Abraham).
- The subjects' task is to formulate the implications of the connection between history and nature into an argumentation that is focused on practical consequences.
- The strategy of the change must be to create power in order to put pressure on the policy makers and the ones they represent in political spheres.
- The goal is to reach a sustainable change through acknowledgment of the consequences of economic acting.

We urgently need a new regime under which the ethical subject thinks and acts independently. This subject should be free of guilt and debt in psychological, religious, and financial spheres. These three aspects are related in deep economy. I have discussed the financial sphere at length. The plea for a radical change in our monetary system, and especially in money's power, will remain impotent as long as the other aspects are neglected. For that reason, I highlight the Jewish conception of guilt and freedom and its parallels with the basic trust in a balance between human history and nature.

In the Jewish tradition, creation and historical liberation are connected in an indissoluble way.[46] The order 'first nature, then history' is self-evident in this vision of life. It concerns the unique way that the Jews handle the question of guilt and liberation, which in many religious and cultural traditions is important in defining the status of the subject. If humanity is free of guilt, nature does not have to do penance. The narrative abundance of God's grace is a daily experience of life in which nature participates. History, morality, and nature as creation cooperate in one movement of liberation, which swallows and annihilates human badness and guilt. One could call this the Jewish sense of life, even in suffering.

There are parallels between the conception of human history in modern enlightenment and in classical Hebrew thought. In both conceptions, history is considered linear; it is a form of progress. The differences are the realization of the progress and the means that are used in its interest. In the Old Testament, God's covenant with Israel is the decisive point. History starts with the people's call for an end to their enslavement and the realization of a free situation for them. Abraham's faith and the Israelites' exodus are two examples. The hope and goal of the Promised Land mean that the only way to reach it is through a promise, a word that denies history's fatal development. The word is always bound to a speaking subject. The Israelites' called this subject YHWH, the name that cannot and should not be pronounced (or appropriated).

This divine power does have representatives on earth. They are never derived from nature. People are always involved; they are members of the promised community. It is important to understand that the people of Israel were already there as a community, but the actual origin of the community was defined by the promise. As soon as faith and morals are detached from the promise, the community loses its raison d'être. As God's representatives, the prophets remind their community of the liberating ground brought forth by the originating promise. The prophets also condemn deviant behavior. The prophets are critical members of the community. This community is not a natural entity—its origin is a promise, not a biospheric cause. The community exists as an 'original' community only when, and as long as, its members realize the call for freedom and are thus faithful to the covenant. As community, people's relatedness to one another and to nature is originated by a divine word. The community is sustained by their common belief that the promise is true. God's promise is history and future development, and it is dependent on the human fulfillment of their part of the covenant.

Israel's prophets speak and hear. They have an understandable message, and this contrasts with the tradition of mysteries and oracles. Instead of vague, enigmatic, ambiguous messages regarding destiny and fate, the prophets deliver plain directions for daily life in the land of freedom. These directions are the conditions for sustaining the covenant relation and thus for making history. The God they represent is not already known, but he is making himself known.[47] The Israelites' faith shows a clear understanding of history, including the continual prophetic or messianic developments and corrections.[48]

The plea for a different sequence—nature first instead of history—is prototypically told in the first chapters of Genesis. These stories are meant to help people know their place. The Hebrews, who already knew about their 'covenant partner' YHWH, told two symbolic, metaphorical stories about the creation of the world and of man and woman. In the first creation story, humans were created on the last day of creation; hence the biosphere was already there. All the creatures, even the dangerous ones, were created by God. All creatures derived their character from the God the Hebrews experienced as a liberator. In the second creation story, humans are supposed to consider nature in general and the beautiful garden in particular as their home. Their lives start as an invitation: Feel at home, use as much as you need. It is as if there is a host trying to make the first people feel comfortable and at home in their new environment: Feel free to use what you need, just as the other creatures do. Every species and creature here takes what is needed (Genesis 1:29–30).

Since everything is already available, humans only have to organize their own living conditions while not disrupting the others' in the process. As Bonhoeffer states, paradise is hope projected back to history's origin.[49] In the second creation story, paradise is the ideal state of the promised land. The origin is the end, from apocalypse to genesis.[50] The situation is not yet here. People will have trouble with the creation of new life as well as with the means of life. The man will have to wrestle with the earth, and the woman will give birth in pain. This is the human condition resulting from their eating the fruit of the tree of knowledge of good and evil, but it is embedded in the abundant availability of nature.

As physical and spiritual beings, people must protect life, and they use nature to do this. Making history means the organization of nurturing one another and creating a culture.[51] For this purpose, people use nature to the extent that they need it.

The historical approach unavoidably leads to some questions: At what point in time did humans start to use more than they needed? When did we begin exploiting nature? When did we start reorganizing things in such a way that our amnesia in regard to the beginning aim of history became dangerously apparent? As our deep economic considerations clarify, the large-scale, systematic 'forgetting' began in the modern Western era. In more general terms, the sequence is the key to answering the question of this evil's origin: The abuse of nature presupposes the abuse of fellow humans. Power and wealth are at stake. The kings are the characters who tell about this primeval human datum. The man in power controls other humans and nature. The Hebrew king, however, served his people as messianic ruler. Nature was included, as Psalm 72 demonstrates:

> The mountains shall bring peace to the people, and little hills, by right-eousness (v. 3).
> He shall come down like rain upon mown grass: as showers that water the earth (v. 6).
> He shall spare the poor and the needy, and shall save the souls [lives] of the needy. He shall redeem their soul [life] from deceit and violence: and precious shall their blood be in his sight (v. 13–4).
> There shall be a handful of corn in the earth upon the top of the mountains; the fruit thereof shall shake like Lebanon: and they of the city shall flourish like grass of the earth (v. 16).

The modern making of history, the conquest and subjection of nature, could be considered human's avenging of nature's enslaving of humans. Yet the Jewish religion and worldview long ago forcefully rejected humans oppressing humans in nature's name, thus indicating and defining the character of nature's dominance over humanity. The biblical conception of history espouses a knowledge of the right balance, our being part of global righteousness between people and nature. The Hebrew narratives are all about promises, warnings, corrections, and directions for a better future and a better history, leading to a perfect balance. The promise's content is decisive for history. It is the concrete land as well as the vision of the peaceful kingdom where species will live together in perfect harmony.

This opposes the Western golden rule of laissez-faire laissez-aller, which functions as the foundational pillar of Western history in the modern era. This principle is a classic example of an ideological use of natural relations: The term 'laissez-faire laissez-aller' has, so to speak, been borrowed from natural relations and developments. In other words, the system is based on the pseudonatural belief that ecology could function as a laissez-faire model to lead and support economy. The special character of the human species, however, does not seem to work that way. A real change of order will show the pseudocharacter of this kind of ideological priority of nature over history. All advocates for natural, biospheric relations as a model for human action should realize that all kinds of models have already been derived from nature, from Aristotle's hierarchical metaphysical structure to modern liberal laissez-faire.

ECOFEMINIST PERSPECTIVES

Ecofeminist thinking exemplifies a sharp, critical analysis of history combined with a positive approach toward nature. Ecofeminism addresses ecology from a feminist viewpoint, and its research offers revealing results. Ecofeminism plays a particularly important role in the discussion about nature and history. I first discuss the views of two important 'classics' and then of three female theologian-philosophers who have recent publications in the field.

Susan Griffin and Carolyn Merchant both offer an original approach to the parallel dualism between man-woman and man-nature in Western history, and they both present a clear perspective on the depths and degrees of the negative vision of the other. Both authors show the unfortunate consequences, culturally as well as socially, that have resulted from the vision of power holding men. They highlight the equality between woman and nature, primarily in the historical sense: Men have suppressed women and nature simultaneously. The coherence between people dealing with one another and with nature is seen in a specific and surprising way.

Susan Griffin's book *Woman and Nature* was an inspirational source and a departure point for many ecofeminists.[52] Griffin specifically describes how, in Western history, nature and women are regarded as enigmatic and dangerous scientific objects that need to be controlled. Science's struggle with 'matter' is described in a collage of indirect citations and expressions about female psyche and male sexuality.[53] Repeatedly, Griffin connects the handling of nature with conceptions of women. For instance, capriciousness is not appreciated; what is not of use is considered a weed. It is a small step from an inferior crop to an inferior human being. The consequences are horrendous: What is considered inferior should 'disappear.'

Using diverse sources of patriarchal thinking, Griffin points out how Western culture is saturated with it. It is found in scientific works, the Bible, other theological documents, handbooks for farmers, books for businessmen, and even etiquette books for young ladies. In opposition, Griffin consistently portrays women as having sources for another type of knowledge expressed through art, mythology, and the long tradition of witches, midwives, female healers, and female prophets.

Griffin clearly shows where in Western history the power lies and how completely opposing voices have been silenced. Yet an increasing consciousness of other possibilities becomes clear. The knowledge of nature's richness flows from a direct connection between humans and nature, and it is verbalized poetically and spiritually. The opposite *mentality* is being accused, rather than the current economic, structural-suppressing action itself.

Carolyn Merchant's *The Death of Nature* is another noted ecofeminist work.[54] To a greater extent than Griffin's work, Merchant's study parallels my line of thought. The rise of a modern philosophy of life in the sixteenth and seventeenth centuries is sketched in connection with the interaction of nature, humanity, and society. There has always been an association between women and nature. Male dominance over women and nature has increased with the rise of a mechanistic worldview. Merchant's interest lies in the historical process behind the emergence of this view and its consequences, both economically and socially. She outlines the meaning of familiar conceptualizations: nature as a feeding mother, nature as an organism related to the ideal society.

Mechanistic philosophy's influence on the exploitation of nature and humans has had enormous effects. On the one hand, all 'disturbances' of the natural order are judged, such as the infamous witch trials. On the other hand, a market system emerges that regulates the intercourse with land to an increasing extent. Merchant asserts that the mechanistic worldview and ideas of power over nature are still dominant in our thought and action. She pleads for cooperation between the environmental movement and the women's movement in order to create new 'social styles.' In her opinion, other economic priorities can be fixed with an egalitarian and holistic worldview.

In both Griffin's and Merchant's books, there is a clear relationship between nature and history. Male-dominated Western history progressed by

using both nature and women. And it is clear that men have a hostile attitude toward the different, the strange, the other. I agree with this ecofeminist
view of women and nature as being suppressed simultaneously. It is a feature of male dominance: Since men hold strong economic and political positions, they are the oppressors.

The weakness in their approach is that although they analyze history thoroughly, nature goes largely unanalyzed, and it is ultimately disconnected from
history in a plea for holism and egalitarianism. Two aspects are intertwined:
Women are identified with nature and are therefore suppressed, and women
indeed stand closer to nature than men do. It follows that women's values
need to be dominant now. This means, in fact, a plea for influence on the
basis of the connection with nature (women being its representatives) so
that history can be saved from exploitation and hostility (men being their
representatives).

Thus the argument closely approaches the deep economic field, but it stops
short; only half the story is being told, and hence part of the problem remains unaddressed. It is one thing to establish the fact of suppressing power
relations, including their psychological and social aspects; it is quite another
to trace the economic relations and ideology that are hidden behind these
aspects. Although I agree that women are more connected to nature than
men, I do not think that women are immune to the economy's moral and
behavioral characteristics. Ecofeminism cannot avoid deep economic analysis, because deep economic realities are part of modern history. If feminists
avoid their role here by considering themselves solely as objects of oppression alongside nature, a new unworkable antagonism will be created, and a
solution will remain out of reach.

The authors' intentions are different. Their reasoning starts from a consciousness of ecological connectedness and the belief that it is women's
place to rescue it. Men disrupted harmonious relationships, and they created
the dualism that is destroying the living environment for all creatures. The
reestablishment of wholeness and harmony is a female task, and men should
follow women in a large-scale process of change. In a way, there is much
affinity between ecofeminism and deep ecology. The feminist contribution
offers a unique approach to social and historical analysis.[55] Deep ecology
does not show a comparable interest in human relations. The resemblance
between the two is found in their approach to nature.

Theological writers certainly react to the tension that has been building
up historically and as a result of modern destructive changes. Anne Primavesi
defines her ecology argument as a consistent refusal to fragment the world
into separate and independently existing parts.[56] And she defines feminism
as a refusal by the original 'other' in hierarchical human society to be reduced to silence or to be in the subordinate 'other' position any longer.[57]

The original other, the woman, will no longer sit silently. Rosemary Radford Ruether follows the same train of thought. In the introduction to *Gaia and God*, she writes:

> Thus ecology, in the expanded sense of a combined socioeconomic and biological science emerged in the last several decades to examine how human misuse of 'nature' is causing pollution of soils, water, and air, and the destruction of plant and animal communities, thereby threatening the base of life upon which the human species itself depends. Deep ecology took this study of ecology to another level. It examined the symbolic, psychological, and ethical patterns of destructive relations of humans with nature. It particularly saw Western culture, sanctified in Christianity, as a major cause of this destructive culture. It explored ways to create a new, more holistic consciousness and culture. *Feminism* also has many dimensions of meaning. As liberal feminism, it seeks equality of women with men in liberal, democratic societies; as socialist feminism, it declares that such equality is not possible without a transformation of social relations of ownership of the means of production and reproduction. Radical feminism declared that the issue was deeper, that we had to look at the patterns of culture and consciousness that sustain male domination over and violence to women. Eco-feminism brings together these two explorations of ecology and feminism, in their full, or deep forms, and explores how male dominion of woman and domination of nature are interconnected, both in cultural ideology and in social structures.[58]

And in the introduction to *The Body of God*, Sallie McFague states:

> *The Body of God* begins with an analysis of the ecological or planetary crisis that we face, suggesting that everyone has a part to play in the planetary agenda, including theologians. My contribution is the model of the body, a model that unites us to everything else on our planet in relationships of interdependence. In my own journey I have discovered the body to be central to Christianity, to feminism, and to ecology. The organic model suggests, I believe, a possible way to rethink humanity's place in the scheme of things: a post-patriarchal, Christian theology for the twenty-first century. . . . The organic model in its classic form . . . because based on the human body, was hierarchical, anthropocentric (as well as androcentric) and universalizing. But another version of the model from sciences (the common creation story) and from feminism suggests a different possibility: a way of thinking of bodily unity and differentiation that stresses a radical interrelationship and interdependence of all bodies as it underscores their radical differences.[59]

The acknowledgment of being different is extremely important. It is empowering to acknowledge difference rather than to have difference imposed by another. The principle of 'difference' differs from the power relations that declare others to be different and accordingly ascribe the others' positions in society and the world. McFague aims at the right relationship, which is based on the unity of different entities. Primavesi rejects being silenced and reduced to the other and offers radical interpretations of biblical texts that were used to serve the ideology of male dominion. In her striking interpretation of creation and Adam and Eve in Genesis, she reveals that dualism is not embedded in text itself but arises from the history of its interpretation (*Wirkungsgeschichte*).[60] She finally arrives at McFague's point of establishing the right relation. And these ecofeminist theologians are right. The right relation between unity and differentiation is urgently needed. Deep economy reveals the false interpretation, ideological swindling, and practical misuse of differences among people and in application to the natural realm. Deep economy teaches us to comprehend the reasons for it on the scale of advantage (profit, loss, inclusion, exclusion, exploitation). To date, ecofeminists have not dealt with these aspects on a deep level, and this includes the aforementioned problems of life, death, and money.

Nonetheless, ecofeminist analysis of unity and differentiation is an important point of departure for research in the field of deep economic structures. As Ruether states, ecofeminism "explores how male dominion of woman and domination of nature are interconnected, both in cultural ideology and in social structures." Ecofeminism correctly asserts that we need an ethical theory of the other that is not captured in the male ideology of general sympathy. This new ethical theory could have the potential to question the self-evident use of virtues that are currently adjusted to commercial society in a 'Smithonian' way.[61]

CHANGE IN ATTITUDE AND VISION

Ecofeminist writers have helped find a perspective to deal with two conditions: division and connection. Division and connection are the two basic options of environmental ethics. We specify them as the deep economic division between people and groups, humans and nature, and as original ecological connectedness. The regaining of ecological connectedness seeks to overcome the divisions by using organic models and ecological concepts of wholeness.

In modern times, people needed the division between themselves and the other and between themselves and nature. They needed this type of extreme individualism as an underlying presupposition for the development of commercial society. The application and interpretation of traditional values and virtues as skills for the individual in economic life by philosophers and economists, such as Adam Smith, made the cooperation between morality and economic acting acceptable.

According to ecofeminist thought, the discrimination against and contempt for women as a lower class and species serve to proclaim women as the others. Because men see women as living closer to nature, women are treated as lesser beings. This is a sharp observation, and the experience is common to many women. Women form one group of discriminated-against persons. As a historical case, the oppression of women is intertwined with the suppression of various classes and races. So the feminist literature I'm dealing with identifies universal themes that are applicable to all victims of oppression.

This oppression's ideological structure has a typical Western shape that is striking in its unity. The process runs as follows: The powerful human male subject breaks off his original connections and relations, consequently proclaiming new relations. These new relations are supposed to work to this subject's advantage. On his powerful terms, this human male subject declares the other to be the other. This otherness differs a lot from the original, metaphysical one, to which philosophers such as Levinas refer. This artificial other is the subject's very own creation: The bourgeois subject creates his own world, including and excluding persons and natural objects— choosing, as if he were a deity. The other has to fulfill the subject's wishes and expectations, and she has to obey his charges. If the other refuses to function this way, she runs the risk of being excluded from all cooperation and relations. Here again, we encounter a deep difference between the historically shaped other and the naturally given other who is there as a challenge. In the first case, the subject shapes the other; in the second, the other shapes the subject. It is the feminists' as well as Levinas's aim to return to the original, metaphysical other. The core question is whether their assessment of the present oppressive situation, missing an analysis of deep economy, is serious enough. Once the other is considered the system's victim, the other's face challenges the better off to be critical subjects, representing the nameless, homeless, and speechless people, as well as nonhuman creatures.

This critical subject's first task is to overcome the delineation of people, cultures, and species as higher or lower on some sort of vertical ruler. The problem, of course, is that the modern economic system is structurally based on these dichotomies.[62] Thus, the organic-holistic approach is proving ineffective, because it does not deal with the origins and results of the system's foundations and daily practices as described in the preceding paragraph. It is not only morality that is to blame for the sustaining of oppressive dichotomies; it is also the economic complex of exploitation and exclusion as a condition for economically profitable cooperation.

We need a moral code to guide us through the process of changing our attitudes and practices. Again, it is helpful to consult Jewish thought, insofar as it focuses on the ontological and social-economic differences of the other (in her metaphysical, original state, as Levinas repeatedly assures us).[63] I discussed the hidden treasures of Western culture in chapter 3. One of the almost forgotten but still solid moral foundations of contemporary social life is the Jewish notion that the other comes first. Indeed, the I's identity depends on the other. This notion prohibits the individual (the self-supporting subject) from claiming and controlling the other's being and subsequently from deciding issues of oneness and division in the subject's own interest. Jewish thought offers the other the opportunity to be himself, herself, or itself, and it gives way to, as McFague defines it, the "bodily unity and differentiation that stresses a radical interrelationship and interdependence of all bodies as it underscores their radical differences."

We can juxtapose elements of the modern Western and Jewish visions concerning the I and the other. The Jewish philosopher Emmanuel Levinas is consulted here, because he works with the tension of freedom and strife between and among all humans, groups, and populations. Levinas asks which will be stronger: the self or the other, self-interest and self-centeredness or openness toward the other? Levinas also discusses the tension between the same and the other.[64] He acknowledges that the original position is gone. Levinas speaks of the other's 'suffering face,' and to my ears, this is contemporary humans' moral challenge.[65]

Modern Western Vision	*Jewish Vision*
I have to free myself.	I am liberated and thus free.
My position is delicate; the other should help me by free will or force.	The other's position is delicate; I should help the other. The other helps me to be a moral agent.
To be free, I'm in need of means to command the other.	The other is needy; I should help the other to be free.
The system promises wealth for its individual successful members.	I am part of a community of promise, and all members should experience the fruits of the promise.
History is progress under the direction of economics and forced cooperation.	History means being set free as a community to achieve the promised land together.
The final situation is the realization of the American Dream for the individual.	The final situation is paradise: the togetherness of all species without consuming each other.

Reorganizing economy and society begins with a change of attitude. This initial change need not be in the dualistic-spiritual sphere, but rather in the sphere of radical interrelationship and interdependence of all bodies, including their radical differences. We must acknowledge and accept interdependence and difference as creational qualifications and not as a consequence of economic structures. The other's 'suffering face' reveals our current social system's offensive neglecting character, and it calls on us to recognize and award the other a full identity. If we are to continue to make history, we must change our attitude toward the other and reorganize our society and economy in such a way that we cease to damage and destroy the other, nature, and ourselves.

NOTES

1. Many aspects, including aesthetic experience and the object-subject debate, are brought together in Hans Robert Jauss, *Aesthetische Erfahrung und Literarische Hermeneutik* (Frankfurt am Main: Suhrkamp Verlag, 1984). For the manipulation of aesthetic experience, see Theodor Adorno, *Aesthetische Theorie*, Gesammelte Schriften, Band 7, (Frankfurt am Main: Suhrkamp Verlag, 1970).
2. The sublime (*das Erhabene*) was an important concept during the Renaissance and early Enlightenment. The concept stems from Plato's ontology and metaphysics of the beautiful (see Jauss, *Aesthetische Erfahrung*, p. 17). In modern times, we see the emancipation of the aesthetic experience from the Platonist metaphysics of the beautiful (Ibid., p. 100).
3. See Eugene C. Hargrove, *Foundations of Environmental Ethics* (Englewood Cliffs, NJ: Prentice-Hall, 1989), pp. 209–14. Jauss speaks of the "satisfaction of pure consumption and kitsch needs" (*Aesthetische Erfahrung*, p. 71).
4. It depends, of course, on aesthetics' meaning. When we identify aesthetics with pure enjoyment and taking pleasure in something, it covers an object's use and utility in the Western world as well as in China (Jauss, *Aesthetische Erfahrung*, p. 71). In the early modern Western world, however, aesthetics was still related to awe and transcendence.
5. See Franz Hinkelammert, *Die ideologische Waffen des Todes: Zur Metaphysik des Kapitalismus* (Münster: Edition Liberación, 1985), pp. 82–104, especially 83. Hinkelammert is discussing Friedrich von Hayek and Milton Friedman's neoclassic economic theory here. This theory changes the economic subject with his or her needs into a subject with arbitrary preferences and "in doing so, changes the definition of economy" (p. 83).
6. See Jauss, *Aesthetische Erfahrung*, pp. 244–92.
7. Ibid., p. 97, refers to the slogan of the 1970s: "The ruling literature is the literature of the ruling class," and he speaks of the "ruling culture industry's hidden interests."
8. Discussing the aspect of *katharsis* (aesthetics' purifying working), Jauss indicates a one-dimensional attitude, which means the enjoyment of the object without any distance or the sentimental enjoyment of the self. He warns of the danger of ideological capturing and manipulated consumption, which means the end of aesthetics' genuine communicative function (Ibid., p. 166).
9. Jauss describes the subject's ability to deny and even to enjoy negative aspects such as the hateful, the terrible, and the injured as the subject's own disengagement, his consciousness of not being part of it (Ibid., pp. 85f.).
10. This is an aspect of the modern problem of alienation. See for example, Ian Barbour, *Ethics in an Age of Technology* (San Francisco: Harper, 1993), pp. 10–5.

11. An extensive study in this field is Alain Corbin, *Le miasme et la jonquille: l'odorat et l'imaginaire social, XVIIIe–XIXe siècles* (Paris: Aubier Montaigne, 1982).
12. There is another aspect to this whole story. In urban and industrialized areas, only wealthy and upper-middle-class citizens can afford to live in relatively 'clean' areas. Millions of people live in heavily polluted environments.
13. Kierkegaard (1813–55), in *The Ethical (Enter-Eller)*, judged that, from his ethical point of view, the aesthetic attitude of life is like no other. The latter he calls hedonistic (pleasure-searching), and the ethical life he considers typical for the Christian existence. Kierkegaard expects the responsible individual to be capable of overcoming hedonism. On this he fiercely opposes Hegel, who considers the individual's existence as amounting to only a small part of the World Spirit (*die Weltgeist*) as the cultural agent. Kierkegaard thus decides on the superiority of the individual decision (the existential *Entscheidung*) concerning the ruling spirit of the age. Kierkegaard is a typical representative of the nineteenth-century focus on the individual as moral agent, deriving his strength from evangelical moral standards. In this view, the individual is to blame for his or her own decisions. In my opinion, however, we have to consider the spirit of the age in order to understand personal choices. This means that we first have to analyze the core of the aesthetic attitude and discover what ethical attitude results from it. We can then try to detect the moment where things went wrong— the moment when people, regardless of all opposing expressions, become blind to pollution and its causes.
14. Ralph Metzner points out the types of transitions in numerous fields that are needed to change this attitude. See his article "The Emerging Ecological Worldview," in *Worldviews and Ecology*, edited by Mary Evelyn Tucker and John A. Grim (London and Toronto: Associated University Presses, 1993), pp. 163–72, especially the scheme on pp. 170–1 (previously published as "The Age of Ecology," *Resurgence* 149 [Nov./Dec. 1991]).
15. Paul R. Ehrlich, *The Population Bomb* (River City, MA: River City Press, 1975); Garrett Hardin, *Exploring New Ethics for Survival: The Voyage of the Spaceship Beagle* (New York: Viking Press, 1968).
16. Franz Hinkelammert, "The Economic Roots of Idolatry: Entrepreneurial Metaphysics," in *The Idols of Death and the God of Life: A Theology*, edited by Pablo Richard et al. (New York: Maryknoll, 1983), pp. 165–93, discusses this Hobbesian theme as the central issue (especially pp. 165–8).
17. In German and Dutch, there is only one word for this whole complex: *Schuld*.
18. On 'guilt and community,' see the classic essay of M. Mauss, *The Gift* (1923–24; London: Henley, 1966). He regrets the falling apart of economy on the one hand (exchange of commodities and services) and of morals and religion on the other (which is called 'symbolical exchange'). Hobbes's contract theory is the basis of guilt; see Thomas Hobbes, *Leviathan*, chap. 14.
19. These three meanings were extensively discussed in chapters 1, 2, and 4 of this book.
20. Michel Foucault uses this term to indicate that power is always in process; he speaks of power's microphysics, as opposed to power's metaphysical character. See Michel Foucault, *Mikrophysik der Macht* (Berlin: Merve Verlag, 1976).
21. The early-nineteenth-century thinker Schleiermacher could not have imagined how his description of religion as 'the feeling of total dependence' (*schlechthinniges Abhängigkeitsgefühl*) would be applied to 'economic religion.'

22. For a good overview of different theories on power, see S. R. Clegg, *Frameworks of Power* (London: Sage Publications, 1989).
23. See Herman E. Daly and John B. Cobb Jr., *For the Common Good: Redirecting the Economy Toward Community, the Environment, and a Sustainable Future* (1989; Boston: Beacon Press, 1994), pp. 437–41; see also Hinkelammert, "Economic Roots of Idolatry," pp. 176–78. About the International Monetary Fund's missions, Hinkelammert writes: "Their requirements are always the same: (1) cut the government's spending on the elderly, the poor and the sick; and (2) destroy labor organization" (p. 177).
24. See Friedrich Nietzsche, *On the Genealogy of Morals* (New York: Vintage Books, 1967); Tracy B. Strong, *Friedrich Nietzsche and the Politics of Transfiguration* (Berkeley: University of California Press, 1988), especially pp. 100–2.
25. This ground rule is disturbed slightly, but not fundamentally, by certain entrepreneurs in rich countries who know how to use their bankruptcies in profitable ways.
26. G. F. Hegel, *Phänomenologie des Geistes* (Phenomenology of Mind), in *Werke*, Bd. 3 (Frankfurt am Main: Suhrkamp, 1970), pp. 342–54.
27. See Rene Girard, *Le bouc émissaire* (The Scapegoat) (Paris: Bernard Grasset, 1982).
28. Manuel G. Velasques, *Business Ethics: Concepts and Cases* (Englewood Cliffs, NJ: Prentice-Hall, 1988), pp. 83–4. The concept of moral or human rights might offer a solution to the problem that before the eighteenth century there was no such thing as a personal right; see Alasdair MacIntyre, *After Virtue: A Study in Moral Theory* (London: Duckworth, 1987), pp. 68–9.
29. On the historical development of rights in the Western world, see Richard Tuck, *Natural Rights Theories: Their Origin and Development* (New York: Cambridge University Press, 1979).
30. See K. S. Shrader Frechette, *Environmental Ethics* (Pacific Grove, CA: Boxwood Press, 1988), pp. 67–81.
31. John Locke, *The Second Treatise of Government* (1690; New York: Macmillan, 1988), p. 17.
32. Karl Marx locates the center of capitalist social relations at this very juncture: Labor has been expropriated and has become a source of surplus value for those who own the means of production. They command not only a piece of nature but also the energies of others.
33. There are parallels with Aristotle's and Adam Smith's concepts of morality. In both cases, morals are derived from nature and natural laws that are metaphysical in character (metaphysics in the sense of totality thinking). Basic rights, such as property, are valid up until the point where they obstruct the system. For this reason, civil rights are often temporarily postponed in certain countries. A well-known ideological reason is 'national security.'
34. For a balanced judgment, see Rosemary Radford Ruether, *Gaia and God: An Ecofeminist Theology of Earth Healing* (San Francisco: HarperCollins, 1992), pp. 218–28.
35. Tuck, in *Natural Rights Theories*, mentions the most important proponents' arguments. See also Joshua Mitchell, *Not by Reason Alone: Religion, History, and Identity in Early Modern Political Thought* (Chicago and London: University of Chicago Press, 1993). He cites Rousseau, who considers "the holy voice of nature stronger than that of the gods" (p. 213).
36. World Commission on Environment and Development, *Our Common Future* (New

York: Oxford University Press, 1987), p. 89: "If large parts of the developing world are to avert economic, social, and environmental catastrophes, it is essential that global economic growth be revitalized."

37. See Thijs de la Court, *Beyond Brundtland* (London: Zed Books, 1990), and *Different Worlds: Environment and Development Beyond the Nineties* (Utrecht: International Books, 1992).

38. See *Our Common Future*, pp. 89–90.

39. This might declare the great success of Francis Fukuyama's writings on liberalism's historical victory; see *The End of History and the Last Man* (New York: Free Press, 1992).

40. See MacIntyre, *After Virtue*, p. 30.

41. See Tuck, *Natural Rights Theories*, pp. 58–81.

42. MacIntyre, *After Virtue*, pp. 25–30.

43. See Thomas S. Kuhn, *The Structure of Scientific Revolutions*, 2d ed. (Chicago: University of Chicago Press, 1970). See also Ian Barbour, *Religion in an Age of Science* (San Francisco: Harper, 1990), pp. 31–6, 51–8.

44. We could summarize Adam Smith's ideological assertions in this way.

45. See Donella H. Meadows, Dennis L. Meadows, and Jorgen Randers, *Beyond the Limits: Confronting Global Collapse; Envisioning a Sustainable Future* (Post Mills, VT: Chelsea Green, 1991).

46. Rosemary Radford Ruether rejects the sequence first liberation, then creation, proposed by some theologians, such as Gerhard von Rad. I suspect that she misses the statement's meaning by not recognizing that the authors of the creation stories knew about God's revelation as a human experience of liberation.

47. This is the core of Karl Barth's so-called revelation theology.

48. From this perspective, Anne Primavesi rewrites the Genesis stories in *From Apocalypse to Genesis: Ecology, Feminism and Christianity* (Minneapolis: Fortress Press, 1991).

49. In German, *zurückprojizierte Hoffnung*, in Dietrich Bonhoeffer, *Sanctorum Communio* (1931; München: Chr. Kaiserverlag, 1960).

50. See Primavesi, *From Apocalypse to Genesis*, p. 2: "Christianity generally looks back to the Day of Creation and forward to the Day of Judgment. Ecology, in its practical approach, faces the present Day of Judgment, Apocalypse now, with faith in a new creation.... For ecologists, apocalypse is not postponed until the end of time, but is a vision of *how things really are* at this moment.

51. See ibid., pp. 238–43.

52. See, for example, Sallie McFague, *The Body of God: An Ecological Theology* (Minneapolis: Fortress Press, 1993), p. 26. She opens the chapter "Cosmology: The Organic Model" with a long quote by Griffin.

53. Susan Griffin, *Woman and Nature: The Roaring Inside Her* (New York: Harper and Row, 1978).

54. Carolyn Merchant, *The Death of Nature: Women, Ecology and the Scientific Revolution* (New York: Harper and Row, 1980).

55. See Murray Bookchin, *The Ecology of Freedom: The Emergence and Dissolution of Hierarchy*, rev. ed. (Montreal: Black Rose Books, 1991), pp. 119–23. He writes, "latent forever in the repressive morality that emerges with patriarchy is a smoldering potentiality for revolt with its explosive rejection of the roles that socialization has instilled in all but the deepest recesses of human subjectivity" (p. 123). Ecofeminism could recognize itself as "a smoldering potentiality for revolt."

56. Primavesi, *From Apocalypse to Genesis*, p. 25.

57. Ibid.

58. Ruether, *Gaia and God*, pp. 1–2.

59. McFague, *The Body of God*, p. x.

60. See Primavesi, *From Apocalypse to Genesis*, chaps. 10–12.

61. See chapter 1 of this book. Murray Bookchin says it in a different way: "From ethics will emerge rational criteria for evaluating virtue, evil, and freedom, not merely blame, sin, and their penalties. Ethics may try to encompass morality and justify its epistemologies of rule, but it is always vulnerable to the very rational standards it has created to justify domination" (*Ecology of Freedom*, p. 123).

62. The problem is very complex indeed; the classical philosophers of the social contract, such as Hobbes and Rousseau, principally assumed human equality. See Mitchell, *Not by Reason Alone*, pp. 52, 144–5.

63. I do not have in mind an absolute exclusiveness of Jewish thought. A prominent Christian theologian, Dietrich Bonhoeffer, was working with the same approach to the concept of the other. In his *Letters and Papers from Prison* (New York: Macmillan, 1971), pp. 381–2, he interprets Jesus' transcendence as 'being there for others.' He simultaneously advocates, however, a rethinking of the relation between the Old and New Testaments, thus indicating Christianity's profound dependence on the original Old Testament, Jewish concepts and God-talk.

64. See Emmanuel Levinas, *Totality and Infinity: An Essay on Exteriority* (Pittsburgh, PA: Duquesne University Press, 1969). Levinas describes the Western ontological vision of freedom, which is opposed to his concept of 'the metaphysics of the other's face': "It is hence not a relation with the other as such but the reduction of the other to the same. Such is the definition of freedom: to maintain oneself against the other, despite every relation with the other to ensure the autarchy of the I. Thematization and conceptualization, which moreover are inseparable, are not peace with the other but suppression or possession of the other. For possession affirms the other, but within a negation of its independence. 'I think' comes down to 'I can'—to an appropriation of what is, to an exploitation of reality" (pp. 45–6)

65. See Levinas, *Totality and Infinity*, pp. 50–1: "The way in which the other presents himself, exceeding *the idea of the other in me*, we here name face. . . . The face of the Other at each moment destroys and overflows the plastic image it leaves me, the idea existing to my own measure and to the measure of its *ideatum*— the adequate idea."

EPILOGUE

FALSE CONNECTIONS—THAT'S WHAT this book is all about. Or should I say wrong and right connections? I have attempted to display the diverse efforts made by classical and modern theological and philosopical authors who claim a 'reality of connectedness.' This concept offers possibilities for the proposition of social principles and moral standards. Most positive writings on connectedness in the areas of nature and creation are meant to tame and ban wrong connections and criticize society's atomistic structure.

The postmodern proclamation of the 'death of man,' after Nietzsche's 'death of God,'[1] questions ontological-moral claims such as these. According to postmodern thinkers, the postmodern individual lives in a fragmented world in which the concepts of humanity, community, and God are virtually nonexistent in everyday life. The modern era is characterized by its stress on human autonomy and dominance over nature.[2] In this context, it is even possible to speak of moral rationality and rational morality. However, bound to the gravitational laws of the economy, the modern subject is set loose on the moral field armed with well-intentioned advice: Adjust your moral principles and standards to the 'natural laws' of market economics.

Postmodern philosophers, highlighting the individual's loneliness and the disappearance of the modern concept of the free subject, do not really focus on the economy's role. They attempt to prove the free subject's disappearence altogether. Concentrating on aspects such as language, power, mimesis, and alienation, they leave the economy's religious character aside. Contrary to my effort to free humanity through analysis of our money-oriented system as a pseudoreligion, many advocates of a postmodern approach to current Western culture are trying to rid the modern agent of his or her powerful freedom.

I refuse to let this happen, but it is hard work to uphold this position. It is one thing to claim that modern people, living in a democratic society, are licensed to choose ('feel free to make the choice of your heart'). It is quite a different problem to know what one is choosing and why, to be aware of large-scale psychological manipulation, and to resist it.

Our predicament, caused by living by the grace of wrong connections, provokes the age-old question, What should we do to improve the world? rather than the current postmodern advice to stop doing whatever it is we do to improve the world. In other words, there is the awakening of the subject's imagination versus its being put to sleep forever. Once we accept the former question, we have to look for the answer. Our first task might be accurately phrasing the question What should we do?[3] This asks for a sharp awareness of the question's who and what. To be quite concrete, sticking to the acts of writing and reading this very book, the author and the reader of this text should be considered the 'we,' being involved in a process of communication of writing and reading in which they feel connected, a process of mutual recognition. Although there is neither synchronism nor personal acquaintance, language and tradition enable the author to share his critical thought with readers, which makes both parties subjects. Readers are challenged to share the writer's thought and develop their ideas and concepts in a critical discussion. All this is an important form of action, which might eventually lead to action in the practical fields of policy making and politics on micro, meso, and macro social levels.

In this book, this first type of action is presented with a system analysis from the victim's standpoint. It is the painful learning process of critical rethinking and of the revelation of a hidden truth. To be quite personal: You and I, reader and author of this text, supposedly are children of the enlightenment. Our connection is rooted in modernity.

Rather than submitting to a postmodern diagnosis of ourselves in a fragmented, speechless, and storyless society, we should incite, as I tried to make clear, a new wave of enlightenment. This wave of enlightenment includes a process of research, of teaching and learning, connecting critical thought and social relations. All premodern, preenlightenment organic models are unfit for this task, including the established institutional religions, such as the Vatican variation of Roman Catholic doctrine and morality or other kinds of fundamentalism. They obscure truth rather than reveal it.

I commenced this book with the metaphor of the Trinity. The doctrine of the Trinity has been functioning from the outset of the history of Christian culture as the symbolic incentive for sociality.[4] Ideologically, the political systems of the Roman Empire and subsequent empires had an interest in the doctrine of the Trinity, including its precise formulation and its general acceptance.[5] Thus, the Trinity's meaning and influence are not limited to personal and religious relations. The profundity of the consequences for humans in power, being considered God's representatives on earth, is as clear as can be.

The concept of dualism may be helpful to understanding the connection of throne and altar. That connection is directly related to the powerful impact of premodern political theology, which resulted in a clear sanctification and legitimation of established, traditional power relations.[6] People's beliefs are involved, and that is all an emperor wants in order to maintain his empire. But there is more at stake. As I have attempted to reveal, a close connection between the Greek and Jewish traditions occurred in the early stages of Christendom and the Christianizing of the Roman Empire. The dualism that originated in this intercourse allowed God's role in the direct contact between God and humans to be replaced by the church and its offices.

The concept of the one historical God who reveals himself or herself in different modes in different historical situations has been transfigured into a multiapplicable, metaphysical concept of God.[7]

Thus, the religious legitimation of wrong connections was conceded, which resulted in far-reaching consequences in social and political life. In this book I concentrated on two major steps. The first political one contains the Christianization of the ancient Greek polis model. The second economic step is the definitive enchantment of the economy by allowing *chrematistics* (moneymaking) to enter the proscenium. The element that Aristotle had excluded as a dangerous way of dealing was introduced as commercial society's core element.

This development presupposes a deepening of dualism. The human subject became disconnected from nature. All mythological remnants of the relation between humans and nature were proclaimed irrational and hence without value. A new connection was attained: human reason being entitled and enabled to reorganize nature for human well-being and thus for general human profit. This is what accounted for the excitement of the so-called second wave of enlightenment, claiming wide, admirable human efforts and successes.

Since the economy's functioning wholly depends on political and moral agreement,[8] the metaphor of the Trinity is excellent, because of the organic connectedness that this image suggests. Reticence against Marxist models of social antagonism and class struggle is an especially fruitful ground for organic ideas, which often border fascism.[9] Advocates of the organic system bitterly blame the critics for disturbing the subtle, secret processes of moneymaking,[10] which I compared to the fertilization that leads to new life.

In this controversy, a third party has recently come into existence, proclaiming a philosophy of connectedness. There is, of course, no question of one party with one philosophy. Despite the many differences, however, all of them oppose the human destruction of natural communities. Although they witness a profound truth of natural connectedness and mutual dependence of humans and other species, deep economic structures are not challenged, due to a genuine hermeneutical problem: how to relate critically the ecological (natural) and the economic (artificial) connectedness. In a morally pluralistic society, ethics that belong to one area (e.g., protection of the environment) cannot automatically be applied to the other (e.g., pollution through production).

A convincing analysis should be embedded in two different areas: the field of reason and that of belief. My introduction asserts that the traditional God-talk did not suddenly disappear without leaving a trace. Increasingly, people acknowledge that we need a God-concept as a 'common good,' as a point of view and orientation that enable the partners in the discussion to talk about their being deeply concerned. It has always been the Spirit's character to inspire and to unite people to do some dirty work, cleaning the traditional symbol's meaning and impact of age-old ideological interpretations, as well as comparing our social system's structure with the traditional triune Holy Name. Call it an ironic proposal to indicate our predicament's seriousness.

Putting this 'dirty work' in the hermeneutical context, a common image of the world's current situation as a global commercial society is requested, as well as common knowledge of our traditional roots. As many thinkers have put it: The hermeneutical circle keeps going by suspicion.[11] Deep economic analysis is such an active moment of suspicion, activating the critical impact of 'God-talk' and 'God-belief.' To get an idea of what this means, a reliable criterion should be applied. Reactions to stimuli can provide such a criterion. A historical example should make clear what I am trying to say. How did the churches and the so-called Christian cultures react to the Holocaust? Dietrich Bonhoeffer, being one of the few German Christians who reacted in any critical manner, claimed as early as 1934 that only people who were crying for the Jews could be considered members of the church.[12] This is all about clearly indicating a concrete, effective religious belief. The God question should not be put on an abstract level, but in this very concrete field of action: What kind of (belief in) God do people and institutions claim, reacting or refusing to react to social injustice?

For this reason, I join Levinas's 'philosophy of reaction,' which includes the self's openness to all that is excluded. The face of the other is asking for a reaction, which means connection. This kind of connection is not a self-evident, natural, or creational state of affairs. Rather, it is the consequence of a moral decision that people make, essentially rooted in some kind of belief.

All this implies a profound critique of the institutional churches. Hardly any of the established churches spoke for the Jews, nor did they criticize deep economy as a 'believer's reaction' to evident macrolevel injustice. This does not mean, however, that these institutions should be excluded from critical action. They should be used as starting points, as realms where God-talk is still practiced. I would advocate, however, a broad ecclesiastical concept. People who want to react to deep economic magic are simply in need of fellow 'antibelievers.' The character of the communities they constitute is that of a person who takes on an openness to the face of the other. Or, put differently: both a personal and a communal conversion from within. What does this mean in everyday practice? Can the subject in fact be saved by a divine miracle? Will Mother Earth, including her human inhabitants— who, after all, are responsible for her potential premature death—have a chance at survival in the end?[13] The only way to go, in my view, is to take the road of life-feeding connectedness.

In this field, two influencial thinkers made major contributions on the subject of the threesome 'I, the other, and God': Emmanuel Levinas and Dietrich Bonhoeffer. Both thinkers suggest that the transcendental experience of God is in attention to and contact with the other. To put it theologically, the incarnation as concept becomes incarnation in practice. God reveals himself or herself in the other and in social relations. This always implies a critical-moral approach to oppressive and exclusive relations. Both Levinas and Bonhoeffer reached their practical-radical standpoint through their experiences with the Nazi ideology and customs.

This attitude conveys that the postmodern 'death of the human,' the annihilation of the freely acting subject, cannot be accepted. A resurrection of the imagined dead occurs in the thought of both philosophers. This concerns a crucial emphasis, a matter of life and death. Resurrection in the theological sense, of course, always ensues in reaction to a disaster, when all hope seems to have faded. My plea for a reappreciation of the active subject rests on these core elements in Bonhoeffer's and Levinas's thought.

My initial aim is therefore not to—theoretically—counter the postmodern ostracizing of the subject with another opinion. Postmodern philosophers who ostracize the subject react to our century's absolute moral low points. Their reliance on Nietzsche denotes the prophetic character of Nietzsche's culture criticism more than anything else. To be sure, I allude to large-scale totalitarian attempts to silence the troublesome, autonomous subject once and for all because the so-called state interests at stake are too high. In addition to the capitalist economy, I have in mind the two disastrous totalitarian regimes: fascism, specifically the Nazi variant, and communism, especially the Stalinist example.

One could say that the Enlightenment ideals, signifying the highest form of human civilization, got jammed in the technology of production and in the perverted perfection of the total control of mass killing.[14] I therefore petition for a paradigm shift toward a new kind of enlightenment, wherein the critical power of the subject really appears. The subject then rejects both totalitatarian control and complete psychological manipulation.

The use of the metaphor of the perverted Trinity is justified by the profound religious character of our economic-political system. The current prevalent Trinity of deadly relations has come to replace the original Trinitarian concept of life-giving communication. There can be no return to the premodern Trinitarian concept; too much stands in the way historically. In the Eastern Orthodox Church, social ethics are still derived from supposed undamaged Trinitarian relations. Hence, a frightened reticence concerning societal relations and the questions that arise from them prevails in Eastern Orthodoxy.[15]

We are challenged to choose a new path, and thinkers such as Bonhoeffer and Levinas can lead the way. In terms of a new Trinity, we can speak of a new incarnation, a messianic experiment, a movement that resembles the exodus from ancient Egypt, the house of slavery. Slaves, doomed and imagined dead, become living subjects.

After the chastening of the deep economic analysis, we now need to consider practical resistance. The Dutchman Willem Hoogendijk and the German Ulrich Duchrow have both made impressive efforts to substantiate a wide range of initiatives to tame the market economy.[16] Yet there is more to note about the nature of acting agents and the nature of the proposed and morally exigent activities.

In the final conclusion of this book, several points need tangible emphasis. Acting is most urgent in fields where deep economic relations are most transparent, where antagonisms between sexes, races, classes, and species clearly occur. An example of each aspect follows.

Women organize in alternative gender-specific institutions, for instance, to put on trial the predominant hostility toward women in the Roman Catholic Church. In doing so, it appears crucial for them to dialectically connect their objections to the societal relations that are maintained by economics and politics.[17] Church and society cannot be separated fully, but surely we cannot equalize them either. Still, there is a limpid institutional reflection, including personal relations and the ideology that sustains these relations.

Concerning the segregation of races, churches and other social groups whose aim is to overcome obstinate racial opposition may want to take a close look at Glide Memorial United Methodist Church in San Francisco. Here, societal analysis and actual social work are deeply connected to the life and affairs of the multiracial community. Real experiences of people who are psychologically and socially in trouble have an important place at Glide, both in its Sunday morning 'celebrations of life' and in the work that volunteers of all kinds, colors, and creeds do on an everyday basis for the struggling community. Reverend Cecil Williams, himself an original thinker and a prominent Democratic Party member, leads the way in establishing a direct connection between social and psychological problems on the one hand and economic and political structures and relations on the other.[18] Under the heading 'walk that walk,' this essential link is evident in a powerful, soulful, and constructive manner.

The old concept of class struggle seems hard to overcome in any ideological fashion. We can, however, focus on the contradictions between poor and rich on the local and global levels. In view of the multiple roles of money that I discussed extensively, we should consider the original initiative of LETS (local exchange and trade systems).[19] In this rapidly growing movement, which branched into international networks in the 1990s, money regains its original function: the exchange of commodities and services. Actual money is replaced with a point system, employing moneylike symbols. Although we should not underestimate the practical workings here, the symbolic significance is even more important. Participants in LETS demonstrate that the imaginative subject can be considered neither dead nor unemancipated. Here we see a grassroots movement that manages to actually touch a nerve of the economic-political system. The continuous development of a critical theoretical framework is essential to avoid falling into the trap of a sectarian, utopian, worldless group. What has been understood theoretically is now substantiated in practice: Harmful relations are initiated and continued through the necessities of money creation and capital accumulation, and actual alternatives need to and can be provided through original thought and cooperation.

Significantly, LETS are propagated in the environmental movement also. Groups in the environmental and ecological field need to divert from the repeated moral-ontological imperative of ecological-natural relations, which does not gauge the economy's structures in depth. Fortunately, environmental problems receive ongoing attention in many institutions such as churches, schools, universities, and politics as a whole. Combined with groups attending to nature traditionally, a powerful social movement has transpired. The question remains, however, whether this 'environmental collective' can fulfill what it lays claim to. Specifically, on the ideological platform, serious discord exists, despite the common emphasis on deep ecological system analysis. The church, with its input concerning a vision of creation from a liberation theology perspective, might prove to be a helpful catalyst. But we need to deconstruct the myth of good connectedness, concerning both deep ecology and deep economy. The essential quality of the concept of justice will be obvious to all who followed the argument.

Thus we see *post*-postmodern agents at work everywhere. We see people who are affected, who feel touched by the ancient tale of the Israelites' God either directly or indirectly through handed-down values and critical standards. This new subject is more modest than the embellished subject of the European Enlightenment, created through its ideology of world peace and a universal moral. These concepts have been betrayed, or perhaps we should say that they were conceived in order to eventually be betrayed. Therefore, they are rendered out of date, and no one cries for them, except maybe an old humanist here and there. We now see that ideals and guidelines from above, call it God or call it the common good, simply cannot work unless they serve to legitimize group interests.

The contours of a new wave of enlightenment are being drawn by people walking the reverse walk—no more insinuating of fancy ideas as the final truth. The point is to work upward from the base. The scheme of base superstructure can hence be activated once more, or rather, finally be put into operation as Marx intended it. Marx's originality was that he completed the scheme and applied it to nineteenth-century England. Yet the scheme itself was derived from his own tradition: Israel's tradition; that is, where even the existence and methods of God (YHWH) were always coupled with the interests of the weakest—the 'option for the poor.' No other than Jesus himself worked with the same conceptions. It is now up to us to once more fill this scheme's content. On both the theoretical and the practical levels, this book attempts to contribute to that process.

NOTES

1. See Richard Kearney, *The Wake of Imagination: Towards a Postmodern Culture* (Minneapolis: University of Minnesota Press, 1988). Identifying the modern subject's power as the freedom of imagination, Kearny presents renowned French philosophers as the subject's gravediggers. Cf. the following qualifications: Lacan: 'the dismantling of the imaginary'; Althusser: 'the imaginary of false consciousness'; Foucault: 'the end of man'; Barthes: 'the death of the authorial imagination'; and Derrida: 'mime without end' (pp. 252–95). Dealing with Roland Barthes, Kearney states that "the postmodern death of the author . . . follows from the death of God and announces that of Man" (p. 276).
2. Cf. Susan Griffin, discussed in chapter 5.
3. This is precisely what V. I. Lenin questions in his famous 1902 pamphlet.
4. See Sigurd Bergmann, *Geist, der Natur befreit: Die trinitarische Kosmologie Gregors von Nazianz im Horizont einer ökologischen Theologie der Befreiung* (Mainz: Matthias Grünewald Verlag, 1995).
5. See Bergmann, *Geist*, and Ulrich Duchrow, *Alternatives to Global Capitalism: Drawn from Biblical History, Designed for Political Action* (Utrecht: International Books, 1995).
6. The term 'political theology' is rather problematic in character. In the twentieth century, the traditional meaning of connection of throne and altar radically altered, becoming a critique on theology's legitimizing of political power structures. Dietrich Bonhoeffer was among the first theological thinkers who applied that criticism to both the highly theoretical and the practical levels, especially in his *Ethics* (1947; New York: Macmillan, 1976) and *Letters and Papers from Prison* (1951; New York: Macmillan, 1971). Opposite to Bonhoeffer stands Carl Schmitt, with his *Political Theology: Four Chapters on the Concept of Sovereignty* (1922; Cambridge, MA, and London: MIT Press, 1985) and *The Concept of the Political* (1932; Chicago and London: University of Chicago Press, 1996). A current cooperative research project on American fundamentalism in the perspective of European political theology, involving the Theology Department of the University of Nijmegen and San Francisco State University's Political Science Department, attends to the thought of both Bonhoeffer and Schmitt.
7. Not quite grasping this hermeneutically tricky field, Jack Miles feeds a tragic misunderstanding, portraying the biblical God YHWH as a divided God. See his *God: A Biography* (New York: Vintage Books, 1995), especially pp. 406–8. Miles, fascinatingly portraying Hamlet as the tragic, divided modern hero (pp. 8–10, 20, 171–2, 397–408), too easily identifies YHWH as the divided 'cult hero.'
8. Most rational ethicists miss this point, for instance, David Gauthier, *Morals by Agreement* (Oxford: Clarendon Press, 1986), especially pp. 83–112.

9. In this context, the 1931 papal encyclical *Quadragesimo Anno* is interesting, including the discussions it provoked. See Richard L. Camp, *The Papal Ideology of Social Reform* (Leiden: E. J. Brill, 1969), pp. 36–40, 144–9.
10. See, for example, Michael Novak, *The Spirit of Democratic Capitalism* (New York: Touchstone, 1982), pp. 237–360.
11. See, for example, Juan Luis Segundo, *The Liberation of Theology* (Maryknoll, NY: Orbis Books, 1977), pp. 7–9.
12. On the complex question of the *status confessionis*, both in Bonhoeffer's context of Nazi Germany and in today's, see Ulrich Duchrow, *Global Economy: A Confessional Issue for the Churches?* (Geneva: WCC Publications, 1987).
13. See Al Gore, *Earth in the Balance: Ecology and the Human Spirit* (Boston, New York, and London: Houghton Mifflin, 1992), especially chap. 13. Interesting in this context are his remarks on the U.S. tradition of the social gospel and the links between social injustice and environmental degradation in today's world (pp. 246–8).
14. The complete depersonalization of the enemy results in the total absence of any feeling of guilt or repentance. Since the perpetrator cannot know the victim personally, there is no question of hate either. These and other related aspects of mass killing have been extensively and impressively commented on by the Dutch author Harry Mulisch, *De Zaak 40–6: Een Reportage* (The Case 40–61: A Report) (Amsterdam: De Bezige Bij, 1988), whose publication in English was wrongly announced in 1998. Mulisch reflects philosophically on the 1961 Jerusalem trial of war criminal Adolf Eichmann, akin to Hannah Arendt's account *Eichmann in Jerusalem: A Report on the Banality of Evil* (1963; New York: Penguin Books, 1992). Arendt, in fact, approvingly refers to Mulisch's book, which was published in German in 1962; see pp. 96–7, 282. Both authors see in Eichmann the ultimate totalitarian example of the mass murderer with no personal hatred toward his victims whatsoever; on the contrary, Eichmann was fascinated, even charmed, by Jewish life, history, and people. Peter Sloterdijk reflected on these matters in his two-volume masterpiece *Kritik der zynischen Vernunft* (Critique of Cynical Reason) (Frankfurt am Main: Suhrkamp, 1983).
15. See John Meyendorff, *Living Tradition* (Crestwood, NY: St. Vladimir's Seminary Press, 1978), and Vigen Guroian, *Incarnate Love: Essays in Orthodox Ethics* (Notre Dame, IN: University of Notre Dame Press, 1987), pp. 117–78. Especially important is Guroian's comparison between Orthodoxy's original social contexts and its Americanization.
16. Willem Hoogendijk, *The Economic Revolution: Towards a Sustainable Future by Freeing the Economy from Money-making* (London: Merlin Press; Utrecht: Jan van Arkel, 1991), and Duchrow, *Alternatives*, pp. 240–315. In Duchrow's work, see especially Bill Moyers's instructive schemes: 'eight stages of the process of social movement success' (pp. 284–5) and 'four roles of activism' (pp. 286–7).
17. See Rosemary Radford Ruether, *Women Church: Theology and Practice of Feminist Liturgical Communities* (San Francisco: Harper and Row, 1985). See also Angela Berlis, Julie Hopkins, Hedwig Meyer-Wilmes, and Caroline Vander Stichele (eds.), *Women Churches: Networking and Reflection in the European Context* (Mainz, Germany: Grünewald; Kampen, the Netherlands: Kok Pharos, 1995). The collected articles are edited in English, German, and French, reflecting the typical European networking problem of the continent's different languages.

18. Cf. Cecil Williams with Rebecca Laird, *No Hiding Place: Empowerment and Recovery for Our Troubled Communities* (San Francisco: Harper San Francisco, 1992).

19. See Duchrow, *Alternatives*, p. 268. Important in initializing alternative trade systems is the Kairos Europa movement (contact address: Hegenichtstrasse 22, D-69124 Heidelberg, Germany). See also R. V. G. Dobson, *Bringing the Economy Home from the Market* (Montreal, New York, and London: Black Rose Books, 1992).

Bibliography

Adorno, Theodor, *Aesthetische Theorie* (Gesammelte Schriften, Band 7, Frankfurt am Main: Suhrkamp Verlag, 1970)

Aeschylus, *Oresteia: Agamemnon*

Arendt, Hannah, *Eichmann in Jerusalem: A Report on the Banality of Evil* (1963; New York: Penguin Books, 1992)

Arendt, Hannah, *The Human Condition* (Chigaco: The University of Chicago Press, 1958)

Aristotle, *Ethics* (Penguin Books, 1976), also: *The Nicomachean Ethics* (London: Loeb Classical Library, 1975)

Aristotle, *The Politics* (Penguin Books, 1988), also: *Politics* (London: Loeb Classical Library, 1972)

Barbour, Ian, *The Gifford Lectures*, vol. 1, *Religion in an Age of Science*; vol. 2, *Ethics in an Age of Technology* (San Francisco: Harper, 1990, 1993).

Bergmann, Sigurd, *Geist, der Natur befreit: Die trinitarische Kosmologie Gregors von Nazianz im Horizont einer ökologischen Theologie der Befreiung* (Mainz: Grünewald, 1995)

Berkhof, H., & Jong, Otto J. de, *De geschiedenis der kerk* (Nijkerk: Callenbach, 1973)

Berlis, Angela; Hopkins, Julie; Meyer-Wilmes, Hedwig; Vander Stichele, Caroline, (eds.), *Women Churches: Networking and Reflection in the European Context* (Mainz, Germany: Grünewald; Kampen, the Netherlands: Kok Pharos, 1995)

Berman, Marshall, *All That Is Solid Melts Into Air: The Experience Of Modernity* (New York: Penguin Books, 1988)

Bloch, Ernst, *Das Prinzip Hoffnung*, 3 vols (Frankfurt: Suhrkamp, 1959)

Bonhoeffer, Dietrich, *Ethics* (1947; London: Fontana Library, 1964)

Bonhoeffer, Dietrich, *Letters and Papers from Prison* (1951; New York: Macmillan, 1971)

Bonhoeffer, Dietrich, *Sanctorum Communio: Eine dogmatische Untersuchung zur Soziologie der Kirche* (1930; München: Kaiser Verlag, 1960).

Bookchin, Murray, *Defending the Earth: A Dialogue Between Murray Bookchin and Dave Foreman*, edited by Steve Chase (Boston: South End Press, 1991)

Bookchin, Murray, *The Ecology of Freedom: The Emergence and Dissolution of Hierarchy*, rev.ed. (Montreal: Black Rose Books, 1991)

Bradford, George, *How Deep Is Deep Ecology? With an Essay-Review on Woman's Freedom* (Novato, CA: Times Change Press, 1989).

Brundtland Report: The World Commission on Environment and Development, *Our Common Future* (Oxford and New York: Oxford University Press, 1987)

Camp, Richard L., *The Papal Ideology of Social Reform* (Leiden: E.J. Brill, 1969)

Capra, Fritjov, *The Turning Point: Science, Society and the Rising Culture* (New York: Simon and Schuster, 1982)

Clegg, S.R., *Frameworks of Power* (London: Sage Publications, 1989).

Corbin, Alain, *Le miasme et la jonquille: l'odorat et l'imaginaire social, XVIIIe-XIXe siècles* (Paris: Aubier Montaigne, 1982)

Cox, Harvey, *The Secular City: Secularization and Urbanization in Theological Perspective* (London: SCM Press, 1965)

Cox, Harvey, *The Seduction of the Spirit* (New York: Simon and Schuster, 1973)

Daly, Herman E., and Cobb, John B., Jr., *For the Common Good: Redirecting the Economy toward Community, the Environment, and a Sustainable Future* (1989; Boston: Beacon Press, 1994)

De la Court, Thijs, *Beyond Brundtland* (London: Zed Books, 1990)

De la Court, Thijs, *Different Worlds: Environment and Development Beyond the Nineties* (Utrecht: International Books, 1992)

Des Jardins, Joseph R., *Environmental Ethics: An Introduction to Environmental Philosophy* (Belmont, CA: Wadsworth, 1993)

Devall, Bill, and Sessions, George, *Deep Ecology: Living as if Nature Mattered* (Salt Lake City: Peregrine Smith Books, 1985)

Dijksterhuis, E.J., *The Mechanization of the World Picture* (1950; London: Oxford University Press, 1969)

Dobson, R.V.G., *Bringing the Economy Home from the Market* (Montreal, New York, and London: Black Rose Books, 1992)

Drewerman, Eugen, *Der Tödliche Fortschritt: Von der Zerstörung der Erde und des Menschen im Erbe des Christentums* (Freiburg: Herder, 1991)

Dubos, Rene, *A God Within* (New York: Charles Scribner, 1972)

Duchrow, Ulrich, *Alternatives to Global Capitalism* (Utrecht: International Books, 1995)

Duchrow, Ulrich, *Christenheit und Weltverantwortung: Traditionsgeschichte und systematische Struktur der Zweireichenlehre* (Stuttgart: Klett-Cotta, 1983)

Duchrow, Ulrich, *Global Economy: A Confessional Issue For the Churches?* (Geneva: WCC Publications, 1987)

Ehrlich, Paul R., *The Population Bomb* (River City, MA: River City Press, 1975)

Eilberg-Schwartz, Howard, *God's Phallus, and Other Problems for Men and Monotheism* (Boston: Beacon Press, 1994)

Elias, Norbert, *Üeber den Prozess der Zivilisation: Soziogenetische und Psychogenetische Untersuchungen*, 2 vols. (1939; Bern: Francke Verlag, 1969)

Euripides, *Ephigenia in Aulis*

Foucault, Michel, *Mikrophysik der Macht* (Berlin: Merve Verlag)

Fromm, Erich, *You Shall Be As Gods: A Radical Interpretation of the Old Testament and Its Tradition* (New York: Fawcett World Library, 1966)

Fukuyama, Francis, *The End of History and the Last Man* (New York: Free Press, 1992)

Fukuyama, Francis, *Trust* (London: Hamish Hamilton, 1995)

Gadamer, Hans-George, Muthos und Wissenschaft, in: *Christlicher Glaube in Moderner Gesellschaft* (Freiburg: Herder, 1982)

Galbraith, John Kenneth, *The Culture of Contentment* (Boston: Houghton Mifflin, 1992)

Gauthier, David, *Morals by Agreement* (Oxford: Clarendon Press, 1986)

Geertz, Clifford, *The Interpretation of Cultures* (New York: Basic, 1973)

Gehlen, Arnold, *Mensch: Seine Natur und seine Stellung in der Welt* (1940; Frankfurt am Main: Klostermann, 1993)

Girard, René, *Le bouc émissaire* (Paris: Bernard Grasset, 1982)

Girard, René, *Mensogne romantique et vérité romanesque* (Paris: Bernard Grasset, 1961)

Gore, Al, *Earth in the Balance: Ecology and the Human Spirit* (Boston: Houghton Mifflin, 1989)

Griffin, Susan, *Woman and Nature: The Roaring Inside Her* (New York: Harper and Row, 1978)

Gupta, J.A., "Women and Health; Fertility, Reproduction and Population," *VENA* 3, no. 2 (Nov.1991), pp.17-21

Guroian, Vigen, *Incarnate Love: Essays on Orthodox Ethics* (Notre Dame: University of Notre Dame Press, 1987)

Guthrie, W.K., *A History of Greek Philosophy* (Cambridge and London: Cambridge University Press, 1978)

Habermas, Jürgen, *Theorie des Kommunikativen Handelns*, 2 vols. (Frankfurt am Main: Suhrkamp, 1981)

Hardin, Garrett, *Exploring New Ethics For Survival: The Voyage of the Spaceship Beagle* (New York: Viking Press, 1968)

Hargrove, Eugene C., *Foundations of Environmental Ethics* (Englewood Cliffs NJ: Prentice-Hall, 1989)

Hartmann, Betsy, *Reproductive Rights and Wrongs: The Global Politics of Population Control and Contraceptive Choice* (New York: Harper and Row, 1987)

Hegel, G.F., *Phänomenologie des Geistes*, in: *Werke*, Bd.3 (Frankfurt am Main: Suhrkamp, 1970)

Heidegger, Martin, *Nietzsche*, Vol. 1, *The Will to Power as Art*; Vol. 2, *The Eternal Recurrence of the Same* (1961; San Franciso: HarperCollins, 1991)

Hinkelammert, Franz, *Der Glaube Abrahams und der Oedipus des Westens: Opfermythen im Christlichen Abendland* (Münster: Edition Liberación, 1989)

Hinkelammert, Franz, *Die ideologische Waffen des Todes: Zur Metaphysik der Kapitalismus* (Münster: Edition Liberación, 1985)

Hinkelammert, Franz, "The Economic Roots of Idolatry: Entrepreneurial Metaphysics", in *The Idols of Death and the God of Life: A Theology*, edited by Pablo Richard et al. (New York: Maryknoll, 1983)

Hobbes, Thomas, *Leviathan, Or the Matter, Forme and Power of a Common-*

wealth Ecclesiastical and Civil (1651; Oxford: Basil Blackwell, n.d.)

Hoogendijk, Willem, *The Economic Revolution: Towards a Sustainable Future by Freeing the Economy from Money-making* (London: Merlin Press; Utrecht: Jan van Arkel, 1991)

Hoogstraten, Hans D. van, "Ethics and the Problem of Metaphysics," in *Theology and the Practice of Responsibility: Essays ons Dietrich Bonhoeffer*, edited by Wayne Whitson Floyd Jr., and Charles Marsh (Valley Forge, Pennsylvania: Trinity Press International, 1994), pp. 223-237

Hoogstraten, Hans D. van, "Europe as Heritage: Christian Occident or Divided Continent?", in *Bonhoeffer's Ethics: Old Europe and New Frontiers*, edited by G.Carter et al. (Kampen, the Netherlands: Kok Pharos, 1991), pp.97-111

Jaspers, Karl, *Plato and Augustine*, (San Diego: Harcourt Brace Jovanovich, 1962)

Jauss, Hans Robert, *Aesthetische Erfahrung und Literarische Hermeneutik* (Frankfurt am Main: Suhrkamp Verlag, 1984)

Jay, Martin, *The Dialectical Imagination* (Boston: Little, Brown, 1973)

John Paul II, Pope, *Centesimus Annus* (1992)

Kearney, Richard, *The Wake of Imagination: Towards a Postmodern Culture* (Minneapolis: University of Minnesota Press, 1988)

Kettenacker, Lothar, "Der Mythos vom Reich", in Karl Heinz Bohrer, *Mythos und Moderne: Begriff und Bild einer Rekonstruktion* (Frankfurt: Suhrkamp, 1983), pp. 261-89

Kierkegaard, Sören, *Either-Or: A fragment of Life* (1843; Dutch edition: *Of/Of: een levensfragment, uitgegeven door Victor Eremita*, Amsterdam: Boom, 2000)

Kuhn, S., *The Structure of Scientific Revolutions*, 2d ed. (Chicago: University of Chicago Press, 1970)

Leeuwen, A.Th. van, *Christianity in Word History: The Meeting of the Faiths of East and West* (London: Edinburgh House Press, 1964)

Leeuwen, Arend Th. Van, *De nacht van het kapitaal: Door het oerwoud van de economie naar de bronnen van de burgerlijke religie* (Nijmegen: SUN, 1985)

Lenin, V.I., *Wat te doen? Brandende kwesties van onze beweging* (1902; Amsterdam: 1976)

Levinas, Emmanuel, *Totality and Infinity: An Essay on Exteriority* (Pittsburg, PA: Duquesne University Press, 1969)

Locke, John, *The Second Treatise of Government* (1690; New York: Macmillan, 1988)

Locke, John, *The Works of John Locke, Including an Essay on the Human Understanding, Four Letters on Toleration, Some Thoughts on Education, and An Essay on the Value of Money*, new ed. (London and New York: Ward, Lock & Co., n.d.)

Löwith, Karl, *Meaning in History* (Chicago and London: University of Chicago Press, 1949)

Löwith, Karl, *Von Hegel zu Nietzsche: Der revolutionaire Bruch im Denken des neunzehnten Jahrhunderts* (1941; Hamburg: Felix Meiner Verlag, 1978)

MacIntyre, Alasdair, *After Virtue: A Study in Moral Theory* (London: Duckworth, 1987)

Marx, Karl, *Das Kapital: Kritik der Politischen Ökonomie* (1867; Berlin: Dietz Verlag, 1973)

Mauss, M., *The Gift* (London: Henley, 1966)

McFague, Sallie, *Models of God: Theology for an Ecological, Nuclear Age* (Philadelphia: Fortress Press, 1989)

McFague, Sallie, *The Body of God: An Ecological Theology* (Minneapolis: Fortress Press, 1993)

Meadows, Donella H., Meadows, Dennis L., and Randers, Jorgen, *Beyond the Limits: Confronting Global Collapse; Envisioning a Sustainable Future* (Post Mills, VT: Chelsea Green, 1991)

Meyendorff, John, *Living Tradition* (Crestwood, N.Y.: St. Vladimir's Seminary Press, 1978)

Metzner, Ralph, "The Emerging Ecological Worldview," in *Worldviews and Ecology*, edited by Mary Evelyn Tucker and John A. Grim (London and Toronto: Associated University Presses, 1993), pp.163-72 (previously published as "The Age of Ecology," *Resurgence* 149 [Nov./Dec. 1991])

Miles, Jack, *God: A Biography* (New York: Vintage Books, 1995)

Mitchell, Joshua, *Not By Reason Alone: Religion, History, and Identity in Early Modern Political Thought* (Chicago: The University of Chicago Press, 1993)

Moltmann, Jürgen, *Trinität und Reich Gottes* (München: Kaiser Verlag, 1980); English: *The Trinity and the Kingdom*, trans. Margareth Kohl (San Francisco: HarperCollins, 1991)

Mönnich, C.W., *Koningsvanen: Latijns-christelijke poëzie tussen Oudheid en Middeleeuwen 300-600* (Baarn: Ambo, 1990)

Mulisch, Harry, *De Zaak 40-6: Een Reportage* (Amsterdam: De Bezige Bij, 1988)

Nagl, Ludwig, "Habermas and Derrida on Reflexivity," in *Enlightenments: Encounters between Critical Theory and Contemporary French Thought* (Kampen, the Netherlands: Kok Pharos, 1993), pp. 61-76

Nietzsche, Friedrich, *On the Genealogy of Morals* (New York: Vintage Books, 1967)

Nilsson, Martin P., *Geschichte der Griechischen Religion*, vol. 1, *Die Religion Griechenlands bis auf die griechische Weltherrschaft*, vol 2, *Die Hellenistische und Römische Zeit* (München: Verlag C.H. Beck, 1967, 1974).

Novak, Michael, *The Spirit of Democratic Capitalism* (New York: Touchstone, 1982)

Paehlke, Robert C., *Environmentalism and the Future of Progressive Politics* (New Haven and London: Yale University Press, 1989)

Pius IX, Pope, *Quadragesimo Anno*, 1931

Postel, E., "Gender, Health and Population Policy," *VENA* 3, no. 2 (Nov.1991): pp.4-7

Potter, Elizabeth, "Gender and Epistemic Negotiation," in Linda Alcoff and Elizabeth Potter, *Feminist Epistemologies* (New York: Routledge, 1993)

Prendergast, Christopher, *The Order of Mimesis: Balzac, Stendhal, Nerval, Flaubert* (Cambridge: Cambridge University Press, 1986)

Primavesi, Anne, *From Apocalypse to Genesis: Ecology, Feminism and Christianity* (Minneapolis: Fortress Press, 1991)

Radford Reuther, Rosemary, *Gaia and God: an Ecofeminist Theology of Earth Healing* (San Francisco: HarperCollins, 1992)

Radford Ruether, Rosemary, *Women Church: Theology and Practice of Feminist Liturgical Communities* (San Francisco: Harper and Row, 1985)

Rawls, John, *A Theory of Justice* (Oxford: Oxford University Press, 1971)

Rawls, John, *Political Liberalism* (New York: Columbia University Press, 1993)

Rich, Bruce, *Mortgaging the Earth. The World Bank, Environmental Empoverishment, and the Crisis of Development* (Boston: Beacon Press, 1994)

Richard, Pablo, 'Biblical Theology of Confrontation with Idols', in *The Idols of Death and the God of Life*, edited by Pablo Richard, et al. (New York: Maryknoll, 1983)

Scharffenorth, Gerta, *Den Glauben ins Leben Ziehen . . . : Studien zu Luthers Theologie* (München: Kaiser Verlag, 1982)

Schmitt, Carl, *The Concept of the Political* (1932; Chicago and London: University of Chicago Press, 1996)

Schmitt, Carl, *Political Theology: Four Chapters on the Concept of Sovereignty* (1922; Cambridge, MA, and London: MIT Press, 1985)

Segundo, Juan Luis, *The Liberation of Theology* (Maryknoll, N.Y.: Orbis Books, 1977)

Shrader Frechette, K.S., *Environmental Ethics* (Pacific Grove, CA: Boxwood Press, 1988)

Sloterdijk Peter, *Kritik der zynischen Vernunft*, 2 vols. (Frankfurt am Main: Suhrkamp, 1983)

Smith, Adam, *The Theory of Moral Sentiments* (1759; Indianapolis, Indiana: Liberty Classics, 1982, photographic reprint of the 1976 Oxford University Press edition)

Smith, Adam, *An Inquiry into the Nature and Causes of the Wealth of Nations*, (1776; Chicago: University of Chicago Press, 1976)

Ste.Croix, G.E.M. de, *The Class Struggle in the Ancient Greek World: From the Ancient Archaic Age to the Arab Conquests* (Ithaca, NY: Cornell University Press, 1981)

Strong, Tracy B., *Friedrich Nietzsche and the Politics of Transfiguration* (Berkeley: University of California Press, 1988)

Tawney, R.H., *Religion and the Rise of Capitalism* (New York: Harcourt Brace, 1926)

Taylor, Charles, *Sources of the Self: The Making of the Modern Identity* (Cambridge: Cambridge University Press, 1992)

Tönnies, Ferdinant, *Community and Society (Gemeinschaft und Gesellschaft)*, with a new introduction by John Samples (1887; New Brunswick, NJ: Transaction Publishers, 1993)

Tuck, Richard, *Natural Rights Theories: Their Origin and Development* (New York: Cambridge University Press, 1979)

Vattimo, Gianni, *The End of Modernity: Nihilism and Hermeneutics in Post-modern Culture* (Baltimore: The John Hopkins University Press, 1988)

Velasques, Manuel G., *Business Ethics: Concepts and Cases* (Englewood Cliffs, N.J.: Prentice Hall, 1988)

Voegelin, Eric, *The New Science of Politics* (1952; Chicago: University of Chicago Press, 1987)

Weber, Max, *The Protestant Ethic and the Spirit of Capitalism* (1905; New York: Charles Scribner's Sons, 1958)

Weber, Samuel 'The Dawn of a New Age,' paper presented at Multimedia Computing Conference, Utrecht, The Netherlands, 1991 (not published)

White, Lynn, Jr., "The Historical Roots of Our Ecological Crisis," in *Science*, 10 March 1967. Republished in Donald VanDeVeer and Christine Pierce (eds.), *Environmental Ethics and Policy Book: Philosophy, Ecology, Economics* (Belmont, CA: Wadsworth, 1994), pp.45-51

Williams, Cecil, and Laird, Rebecca, *No Hiding Place: Empowerment and Recovery for our Troubled Communities* (San Francisco: Harper San Francisco, 1992)

INDEX

accumulation: of capital, 12–3, 22, 24, 26–7, 47, 49–50, 54, 56–7, 99, 118–9, 131, 153; of money, 100, 119, 131; of wealth, 27, 45, 47

aesthetics: attitude, 116–7; and ethics, 114; and mass-media, 115

alienation: and community, 48; and dualism, 79; postmodern philosophers on, 147

anarchism, philosophical: of Bookchin, 57; and ideology, 57

anthropocentrism: and biocentrism, 37–8; and dualism, 74; as an excuse, 131; and religion, 5, 61; social system, 1; Western culture's, 38

Aquinas, Thomas: natural reason, revelation and dualism, 76; Paul E. Sigmund on, 84n. 44

Arendt, Hannah: Eichmann process, 155n. 14; on polis, 31–2n. 35

Aristotle: on chrematistics and money, 11, 12, 45, 46; on community, 9, 10, 13, 44, 45, 46; on cooperation of lord and slave, man and woman, 10, 11, 92, 110n. 19; on enlightenment, 8; on eudaimonia, 9, 10, 11, 12, 68; on natural and unnatural, 9, 10, 11, 12, 13, 46, 69; on natural order, 9, 10; on slavery, 9, 10, 11, 46

attitude: aesthetic, 115–6, 128–9; change of, 143n. 14; critical, 97; egocentric, 98; and population problem, 94; toward life and death, 93–5

Augustine, Saint: the city of God and dualism, 75–6; Jaspers on, 84n. 41; and Plotinus, 84n. 39; Ruether on, 84n. 40

autonomy: and sexual ethics, 103–6

Barbour, Ian: on Deism and evolutionary thought, 30n. 15; on Maslow, 111n. 30; on paradigm shift, xiii, xv n. 6

Barth, Karl, 4

Bergmann, Sigurd: Trinity and cosmology, 29n. 6

Bible: biosphere and, 53; Christian culture, 64; concept of history, 135; covenant, 1; creation, 70; and dualism, 67; God, 2; humanity and nature, 64, 67; Old Testament, 106; patriarchal thinking, 136; prophetic God, 49; radical interpretations, 139; Saint Augustine, reading the, 75; stories, 64, 66; teamwork with nature, 70; working history, 70; worldview, 71

biocentrism: advocates, 56; vs. anthropocentrism, 37–8; equality of life-forms, 39–40; and morals, 56; and power, 56–7

biosphere: advocates, 8; as a community, 51, 133; connectedness, 40; ecology and, xi; equal right, 40; and food chains, 35; God in, 53; as life condition, 108; as model, 135; and moral standards, 9, 51; and perspective, 49; and rights, 125; as a society, 50; universal, 48, 52

body: of God, 138; unity of the, 138, 141

Bonhoeffer, Dietrich: "being there for others," 146n. 63; and Jewish thought, 2; and the Jews, 150; mandates, 112n. 42; and metaphysics, 29n. 8; on paradise, 134; on political theology, 154n. 6; on the relatedness of God and social reality, 29n. 8, 151; on a world come of age, 58n. 7

bonum commune. *See* common good

Bookchin, Murray: anarchism, 57; ethics, 146n. 61; patriarchy, 145n. 55; power, 56; social ecology, 38; social ecology and deep ecology, 56

bourgeois: the autonomous, 6; as God, 6, 140; as *homo economicus*, 46–7;

INDEX

the, individual, 47, 79; and proletarians, 23; society, 47
Bradford, George: biocentrism and anthropocentrism, 37–8; on William Catton, 41; deep ecology, 37–8; power relations, 40–1
Brundtland Report: economic growth, 127, 145n. 36; sustainable development, 127

Calvin, John: theocratic society, 76
capitalism: Daly and Cobb on, 54; industrial, 93; liberal, 127; protestantism and, 28; reflection on, 8; state, 27
chrematistics: enchantment of the economy, 149
Christianity: anthropocentric, 5; love in, 18; scheme of interpretation of, 1–2, 4
church: and progression, 97; and sexual ethics, 103, 104
class: antagonism, 152; social, 20, 47, 54, 79, 124, 126, 140; struggle, 23, 149
Cobb, John, Jr., and Daly, Herman: on biocentric equality, 39, 40, 81n. 10; on biosphere, 51; on chrematistics, 45, 46; on community, 7, 41, 42, 43, 44, 45, 46, 48, 49, 50, 51; on connectedness, 49; and deep ecology, 37, 40, 48, 49, 50; on ecology and economy, 23, 41, 45, 49, 50; on free market economy, 3; on misplaced concreteness, 9, 41, 49, 50; on money, 24, 54, 55, 56; on morality, 51; on politics, 49; on Frederick Soddy, 54
commercial society: and chrematistics, 149; Thomas Hobbes and John Locke on the, 14; and ideology, 150; and individualism, 139; and morals, 24, 125, 139; Adam Smith on, 20, 21; and success, 27, 93
commodity, exchange of, 12; money as a, 15, 16, 20; production and exchange of, 55
common good: accumulation and the, 45; in communities, 91; and economy, 23; ideology of the, 53;

individual interests and, 92; and money, 24; and property, 12; protection of, 14, 125
commonwealth: nutrition and procreation of the, 15; as an organic system, 14
communication: and creation of money, 100; life-giving, 151; multi-racial community, 152; as mutual recognition, 148; and schizothymia, 79; and society, 43; technology, 23
communism, 8, 62, 127, 151
community: as an aim, 10; animal, 138; biophysical, 42; biospheric, 51, 133; Christian, 2; commercial society as a, 21; concept of, 7, 8, 44–6, 48, 51, 56, 99; ecological, 9; and economy, 12–9, 39, 41, 48, 100; enlightened, 120; of equals, 16; Gemeinschaft, 48; and individuals, 17, 21, 42, 44, 57, 99, 108, 109; Israel as, 49–50, 64, 133; just, 42, 57; members, xiii, 16, 20, 50–1, 52–3, 105, 109, 133; moral, 20; national, 39; natural, xv, 9–10, 12, 46, 47, 149; and oikia, 11; old structures, 14; original, 42, 43, 44, 48; and the other, xiv, 16, 53, 150; polis (political), 9–11, 12, 45–6; and postmodernism, 43, 147; premodern, 17, 20, 46, 47, 70–1, 98, 103, 107; promised, 133, 141; relations, 14; small-scale, 91, 107; social, 42; society-community, 48–9; structural conditions for a, 19, 41, 50; as subject, 89; trading, 19; Trinity's, 5
competition: and creation of money, 26, 99–100; and market economy, 45; and the other, 80; and productivity, 130; and society, 55–6, 100, 130
connectedness: Aristotle on, 10, 12; biospherical, 40; in community, 48, 50; cosmic, 61; ecological, 137, 139, 149; economic, 28, 149; life-feeding, 150; with nature, 35, 49, 147, 149; reality of, 147; reality's, xiv; Smith on, 13, 21; of species, 8
connection: division and, 139; new, 149

INDEX

religious character of, 3, 8, 12, 52–3; and Trinity, 21–2, 57; and violence, 96

Deism, 21; God in the individual, 19; God in nature, 13, 19; God as watchmaker, 5, 77; Charles Taylor on, 110n. 21

Descartes, 14

desire: and aesthetics, 115; to be autonomous, 36; collective, 128; Freudian, 95; to be on God's side, 77; and imagination, 117; to kill, 87; object of, 115–6; to possess, 116–7; spiritual, 107; for rights, 124; for wealth, 10

Des Jardins, Joseph: on Bookchin, 56–7

destiny. *See* fate

destruction: and creation of life, 95–9; and creation of money, 95, 96, 108; deep economy and, 96; of life, 96; male, 96; of nature, 96, 108, 118; self-protection and, 108

Devall, Bill: on deep ecology, 6, 39

disenchantment: and facing fate, 89; Max Weber on, 109n. 11

distribution: advancements in, 93; exchange and trade as means of, 12; fair, 24, 26, 102; on a global scale, 89; of *manna*, 74; of power and the economy of salvation, 3; Radford Ruether on, 91; unfair, 119

diversity: of individuals, 48, 51; of life, 38

division: and connection, 139; of converts and former tradition, 64; and guilt, 120; of humans and nature, 5, 9, 67, 71, 72, 139; of I and the other, 139, 141; of knowledge, 41; of labor, 16, 22; of money and power, 95, 101, 102; of morals and economics, 50; of proprietary right and unjust, 125; rulers and ruled, 67; of sexes, 90; in society, 80, 102; of spirit and nature, 69, 71, 73; of subject and object, 75; of things and ideas, 69; three-person, 4

Drewermann, Eugen: original connectedness, 81n. 8

dualism: bible and, 64–6, 139; on rights, 126; and sustainable development, 127; throne and altar, 148. *See also* division; nature

Dubos, Rene: Franciscan and Benedictine debate on nature and environment, 81n. 7

Duchrow, Ulrich: on ancient empires, 31n. 27; on Aquinas, 84n. 42; on market economy, 151

earth, capacity of the, 23, 74, 86, 102, 125; and dualism, 2, 66, 69, 70–1, 73, 75, 79; economy and the, xii; God on, 5; money and fruits of the, 15; Mother Earth, 150; and population, 86, 93, 102, 106; resources of the, 39, 70, 102; "spaceship Earth," 86; and species, 37, 40; subordination to the, 36, 37–8, 40

Earth First, 36, 38

ecofeminism: on dualism, 135, 139; on male dominance, 135, 136–9; on nature and history, 135–9; on the other, 135, 137–40; on women and nature, 135–6

ecology: and Christianity, 145; and feminism, 138–9. *See also* deep ecology

economy: religious character of, 147. *See also* deep economy

egalitarianism: and economic priorities, 136–7

egocentrism: of bourgeois, 98; of parents, 105

egoism: as moral engine, 19; and nature, 19, 43; and original position, 42; and sympathy, 18, 20; and Western *self*, 6

Eilberg-Schwartz, Howard, *God's Phallus*, xv n. 4, 29n. 3

election: in Protestantism, 28, 77; wealth as sign of, 28, 64, 77

Elias, Norbert: on the elite's role in Western civilization, 30n. 20

empire: Roman Empire, 46, 71, 75, 87, 148; two empires doctrine, 71, 75, 76

encyclicals: *Centesimus Annus*, 111n. 40; *Quadragesimo Anno*, 155n. 9

INDEX

Gadamer, Hans-Georg: on a third wave of enlightenment, 82n. 22, 112n. 47
Galbraith, John K., *The Culture of Contentment*, 34n. 82
Gauthier, David, *Morals by Agreement*, 154n. 8
Geertz, Clifford: on religion as fiction and religion as power, xii; on religion as a model, xv n. 4
Girard, Rene: on mimesis, 34n. 83
Glide Memorial United Methodist Church, 152
God: the covenant of, 132–4; as creator, 5, 66–7, 133–4; the Father, 1–6, 57, 77, 85; and liberation, 66–7; the metaphysical concept of, 149; and natural order, 70; as "watchmaker," 4, 5
Gore, Al: on politics and environmental ethics, 30n. 18; on social justice and environmental degradation, 155n. 13
Greek mythology: Aeschylus, 87; Agamemnon, 87; Euripides, 87; fate, 68–9, 87; Helen, 87; Homer, 87; Iliad, 87; Iphigenia, 85–8, 95; Oedipus, 86, 87–8, 95; tragedians, 108
Griffin, Susan: *Woman and Nature*, 135–7
Grotius, Hugo: on proprietary rights and providence, 128
growth: the Brundtland Report on, 127; in commercial society, 21; as ecological problem, 86, 118; economic, 24, 26, 39, 45, 99–100, 119, 121; exponential, 54–5; of the individual's wealth, 27; and innovative processes, 96; necessity of, 57; population, 92, 102, 104; and power, 22; and spending money, 27; as supreme end, 49; victims of, 44
guilt. *See* debt
Gunter, Peter: on man and nature, 7
Guroian, Vigen: on orthodoxy's Americanization, 155n. 15; on Trinity, 29n. 6

Habermas, Jürgen, *Theorie des Kommunikativen Handelns*, 30n. 19

harmony: deep ecology on, 39; of humans and creation, 67; of the natural order, 68–9; Adam Smith on, 18
Hartmann, Betsy: on contraception, 110nn. 12, 16; reproductive rights, xiv, 90, 102
hedonism: Augustine on a culture of, 76; and individual salvation, 6
Hegel, Wilhelm Friedrich: guilt, 120–1; Trinitarian metaphor and philosophy of history, 4, 14
Heidegger, Martin: on dwelling, 35; on nature, 82n. 21
hermeneutical: circle, 150; connection, 95; field, 46; method, xiv; problems, 53, 70, 149; task, 44; tools, 41
Herodotus, 36
Hinkelammert, Franz: on the Abraham tradition and Iphigenia tradition, 85–9; on the economic subject, 142n. 5; on the economists Hayek and Friedman, 142n. 5; on International Monetary Fund, 144n. 23; on sacrificing children, 85–9; on Western culture, 88
history: biblical, 2, 65–7, 72, 73, 76, 132–3, 135, 141; colonial, 78; of interpretation of, 67, 70, 139; modern, 46, 137; nature and, xiv, 9, 126, 128–32, 135–6; philosophy of, 4, 14; salvation in, 72; Western, 4, 7, 37, 57, 61, 67, 75, 78, 97, 135–6, 141; world, 4, 8, 61, 63. *See also* paradigm, shift
Hobbes, Thomas: on money, 14–5; on the mortal god, 14; on power, 14; on the sovereign, 16
holism: in consciousness and culture, 138; economy and, 136; and egalitarianism, 136–7; in nature and history, 137; organic, 140
Homer, *The Iliad*, 87
Hoogendijk, Willem: Western economics and ideology, 31n. 30, 152
humanity: Augustine on, 75–6; and dualism, 76, 78, 79; and economics, 61; Enlightenment and, 108; fatal behavior of contemporary, 3; God

INDEX

and trinitarian formula, 22
legitimation: natural, 22; of power,
148; religious, 149; theoretical, 37;
transcendent, 2
Lessing, Gotthold: on the Trinitarian
metaphor and philosophy of history,
4
LETS (local exchange and trade
systems): on money, 152–3
Levinas, Emanuel: on Jewish thought,
xv, 2; on messianism, 83n. 28; on
ontology and Heidegger, 31n. 28; on
the other, 140, 141, 146nn. 64, 65,
150, 151; on the other's face, 34n.
84, 142; on the other's power, 59n.
18
liberation: experience of, 72; God's,
64–6, 70, 73–4; groups, 104; history
of, 67; individual, 104; from law,
36; of the mind, 75; from nature,
77–8; from power, xii, 1; theology,
xiv, 85, 153
liberty: fate and, 89
life: destruction of, 96. *See also* death
life sustenance: creation of money and,
92; and exploitation and protection,
70, 88; and the population problem,
88
limitations: economic, 26; natural, 37
Locke, John: on money, 14–6; on the
right to private property, 124–5
logos: and *dabar*, 58n. 2; of ecology,
35; and metaphysics, 36; and morals,
36; and mythos, 35, 68; as reason,
35, 69; *spermaticos*, 66, 69
Luther, Martin: two empires of God
and dualism, 76

MacIntyre, Alasdair: on rights, 144
male. *See* power
Malthusian tradition: in environmental
ethics, 86, 88; and violence, 86
market (free market): command
structure, 119; Daly and Cobb on,
42; the divine character of the, 52,
77, 127; economy, 1–4, 6, 7–8, 13,
23, 24, 26, 37, 43, 45, 47, 49, 53,
55, 57, 88, 100, 102, 147, 152; and
Enlightenment, xi; John Locke on,
15–6; Marx on, 22; and moral

capital, 48–9; Adam Smith on, 16,
21, 43
Marx, Karl: on annihilation, 108; on
base and superstructure, 153; on
capital's "canine hunger," 102; on
class struggle, 149; on commodity's
value, 12; on creation of money,
55; and eighteenth-century
Enlightenment, 108; on exploitation,
23; on the "invisible hand," 21; on
labor, 144n. 32; on money, 23, 55–6;
on the origin of money, 22; on
Adam Smith, 4, 21, 108; on
transformation of money into capital,
111n. 34; on the Trinitarian formula,
21
McFague, Sally: on ecology and
feminism, 138–9; on the father
image, 3; on interrelationship and
interdependence, 138, 141; on the
organic model, 111n. 30
Merchant, Carolyn, 135–7; and *The
Death of Nature*, 136
metaphysics: of the Greek community,
9, 13, 46
Miles, Jack: on the divided God, 154n.
7
mimesis: and modern society, 34n. 83
miracles: nature and liberation, 66–7,
70, 72, 74
Mitchell, Joshua: on social contract,
59n. 30
modernity: and aesthetics, 114; and
Christianity, 63; and community, 43;
and dualism, 61; and the experience
of death, 98; a flight from, 106; the
rise of bourgeois society, 47
Moltmann, Jürgen: on Trinity, 30nn.
12, 13, 17
money: functions of, 99; and
sustainable development, 127; theory
of economy, 131. *See also* creation
of money
moral: acting, 6, 121; agent, 104, 141;
agreement, 149; Aristotle's thought,
9–12; community, 20; ideal in Stoic
philosophy, 87; philosophy of Adam
Smith, 13, 17–23, 27, 92–3;
rationality, 147; rights, 122–6;
standards, xi, 6, 9, 86, 103, 104,

Index

Israel's, 65, 134–5; monotheistic, 4;
natural, 64; new, 37; politics and,
75; revelation versus, 25; traditional,
2; Western, 62
representation: divine, 1, 16, 89, 133;
of God, 2, 5, 64, 65, 116, 133, 148;
of God's people, 70; of the human
being, 79; of the metaphysical realm,
104; of the natural order, 10; of
origins and tradition, 1; the other's,
140; of the Spirit, 4; of the sublime,
115
reproduction: means of, 138
responsibility: of the church, 76; of the
free subject, 104; Greek thought on,
46, 69; human, xi, xiii, 19, 104;
personal, 49; private, 26; of society,
48
resurrection: of the imagined death,
151
Rich, Bruce: on the World Bank, 33n.
81
righteousness: biblical, 49, 64, 66,
134–5; and economy, xii; and power,
119; pro-life, 103; the truth of, 49
rights: allocation of, 123, 126; of
animals and natural objects, 62, 126,
civil, 128; environmental, 125; of
future generations, 126; human,
88–9, 123, 124, 125; judicial, 125;
legal, 122–3, 124–5; to live, 124–5;
metaphysical, 144n. 33; moral,
122–3, 125–6; natural, 126, 144n.
35; personal, 122, 126; proprietary,
49, 124–5, 128; reproductive, iv,
90, 102, 104; to live, 125; universal,
126
Roman Catholic: morality, 148;
progression, 97
Ruether, Rosemary Radford: access to
land, 91; on conversion and
catastrophe, 111n. 26; ecology and
feminism, 138–9; on Gnosticism and
Platonism, 83n. 35; linking nature
and covenant, 82n. 18
Ruskin, John: on wealth, 33n. 74

sacrifice: of children, 85–9
salvation: biblical, 72; economy and, 4,
5; individual, 6, 77; Max Weber on, 28

Schmitt, Carl: on political theology,
154n. 6
secularization: and community, 43; and
dualism, 72; and freedom, 58; and
religion, 1, 2; of Western Europe,
77, 106
self: interest, 37, 141; protection and
destruction, 108
semantics: on ecology and economy,
35; in a historical and social context,
44; of language and textual meaning,
44, 71; of text and context, 64; of
the Word of the Lord, 53
Sessions, George: deep ecology, 38, 39
(*see also* Devall)
sexuality: and abortion, 102–3, 105;
and the creation of new life, 102–5;
and dishonesty, 102; and fertility,
65; male, 105, 136; male and female,
110n. 16; and politics, 64, 70; and
population, 85, 92; and power, 79,
103–5; and responsible behavior,
105; and rights, 104; violence and
conception, 95; vulnerable, 103
slavery: Aristotle on, 10–3, 22, 46, 68,
72, 89, 92; the Bible on, 2, 65, 66,
70, 78, 132, 152; in colonial history,
78; and nature, 126
Smith, Adam: on community, 13, 16,
17, 20, 50; on Deism, 13, 19, 30n.
16; on division of labor, 33n. 67; on
economics and morality, 13, 14, 17,
21; on Enlightenment, 8; on the
impartial spectator, 17, 19, 32n. 60;
on the individual, 16, 20, 21; on the
invisible hand, 20, 21; money, 13,
20, 21, 33n. 61; on morality, 4, 13,
50, 57, 139; on natural laws, 19; on
nature and natural, 13, 14, 19; on
prudence and benevolence, 16, 17,
129; on self-love, 18, 20; and Stoic
philosophy, 18; on sympathy, 16, 17,
18, 20, 50; on virtues, 13, 16, 50;
and Western tradition, 9
social: aim, 9; animal, 44, 48; being,
17, 42, 50, 52; body, 25; classes, 10,
20, 47; directives, 3; ecology (*see*
deep ecology); hierarchies, 57, 72;
laws of, system, 22; order, 11, 65,
67, 68, 70, 71; paradigm, xii, 61;

INDEX

reality, xiii, 9, 12, 13, 43, 46, 57;
self, 6; system, 2, 23; theory, 13, 44
social contract, 47
Socrates: reason and nature, 69
Soddy, Frederick: money and debt,
54–5
solvency, 120–1
sovereign: Aristotle on the, community,
10; as God's representative, 116;
Hobbes on the, 15, 16
Spirit: the Church's claim of the Holy,
76; and dualism, 74–5, 78; Hegel's
concept of history and, 14; Holy,
and ethics, 2–5, 57; the human, 46,
57; in the Latin West, 62; as *logos
spermaticos*, 69; and nature, 69;
Nietzsche on the power of, 120; of
pollution, 114
Stoic philosophy: on imperturbability,
87; on interrelatedness, 82n. 24;
spermaticos, 66, 69
subject: bourgeois, 140; free, 147
subsidiarity: in Catholic doctrine, 48
superman (*Uebermensch*): in Athens,
46; Nietzsche, 40
sustainability: of ecology, 130–1; of
economy, 131; of history, 130–1; of
nature, 127, 131
sustainable: change, 132; covenant,
133; development, 10, 24, 88, 126,
127–30; development of economy,
121, 127; future, 39, 41
sustenance: community's, 133; male
dominance's, 138; models of, 8;
morality and economy, 140; of
natural order, 10; as revelation, 74;
of societal relations, 152; system's,
99; women's, 96
symbol: anti, 65; biblical, 66; creation
order as a, 67; dominant, 1; the
gardener as a, 21; God as a, 2; the
King as a, 65, 70; language as, 71;
meaning of, 150; money as, 152;
narrative, 87; nature as, 62, 64, 66;
phallus, 64; religious, 1, 3; signs, 72;
Trinity as a, 1, 3–6, 148
sympathy. *See* virtues

Taylor, Charles: on Deism, 30n. 15,
110n. 21

theology: Bonhoeffer's, xiv; economy
and, 13, 127; liberation theology,
xiv, 85, 153; natural, 13, 36, 40;
political, 3, 148, 154n. 6; and the
population problem, 85; post-
patriarchal, 138
Third World: First, Second, and, 91,
92, 122; and Western power, 101–2,
104, 119
Tönnies, Ferdinand: community and
society, 48
trade: Aristotle on, 11–2; local
exchange and, systems, 152–3;
Locke on, 15–6; and money, 55, 99,
100; Smith on, 21, 26, 57; unions,
100
transubstantiation: proces of, 55, 56
Trinity: applied to economics, politics,
ideology, and ethics, 3; and
cosmology, 29n. 6; Eastern Orthodox
thought on, 151, 155n. 15; and the
economic formula, 21–2, 55; and
incarnational structure, 1; Islam and,
29n. 1; metaphor of, xiv, 148–9,
151; and philosophy of history, 4;
and psychoanalysis, 29n. 3
trust, 50

United Nations and population
problem, 90
utilities: and commodities, 16; demand
for, 26; maximization, 42; outdated,
96; producing and selling, 27;
purchase, 119; trade of, 12; use-
value, 47
utopian: communities, 113n. 57, 153;
thought, 70

value: aesthetic, 114; exchange, 115;
intrinsic and inherent, 114, 125; of
nature, 121; symbolic, 115
Vattimo, Gianni: nihilism, 79, 84n. 51
violence: deep economy and, 96;
feminist thinkers on male, 96
virtues: as basis of commerce, 20, 139;
benevolence, 16, 17; and
connectedness, 50; fairness, 50;
foundation of trust, 43; and free-
market economics, 23; and the
individual, 139; justice, 16, 50; law
of, 6; and natural reason, 76;